Conifers of the New England-Acadian Forest

OTHER BOOKS FROM BRIGHT LEAF

CONIFERS

NEW ENGLAND

ACADIAN

FOREST

A CULTURAL HISTORY

STEVE KEATING

BRIGHT LEAF
BOOKS THAT ILLUMINATE
Amherst and Boston
An imprint of University of Massachusetts Press

Conifers of the New England–Acadian Forest has been supported by the Regional Books Fund, established by donors in 2019 to support the University of Massachusetts Press's Bright Leaf imprint.

Bright Leaf, an imprint of the University of Massachusetts Press, publishes accessible and entertaining books about New England. Highlighting the history, culture, diversity, and environment of the region, Bright Leaf offers readers the tools and inspiration to explore its landmarks and traditions, famous personalities, and distinctive flora and fauna.

ISBN 978-1-62534-787-9 (paper); 788-6 (hardcover)

Designed by Sally Nichols
Set in Adobe Freight Text Pro

Printed and bound by Books International, Inc.
Cover design by adam b. bohannon
Cover art by Charles Edward Faxon c. 1897

Library of Congress Cataloging-in-Publication Data
Names: Keating, Steve, author.
Title: Conifers of the New England-acadian forest : a cultural history /
Steve Keating.
Description: Amherst : Bright Leaf, [2024] | "An imprint of University of
Massachusetts Press." | Includes bibliographical references and index. |
Identifiers: LCCN 2023046464 (print) | LCCN 2023046465 (ebook) | ISBN
9781625347879 (paper) | ISBN 9781625347886 (hardcover) | ISBN
9781685750626 (ebook)
Subjects: LCSH: Conifers—North America. | Evergreens—North America. |
Forest ecology—North America.
Classification: LCC QK494 .K43 2024 (print) | LCC QK494 (ebook) | DDC
585.097—dc23/eng/20240123
LC record available at https://lccn.loc.gov/2023046464
LC ebook record available at https://lccn.loc.gov/2023046465

British Library Cataloguing-in-Publication Data
A catalog record for this book is available from the British Library.

To my parents,
Tom and Jennie Keating,
who first drove me to Maine
many years ago

Contents

Illustrations

FIGURES

TABLE

Acknowledgments

I WOULD LIKE TO THANK everyone at University of Massachusetts Press who played a role in making this book possible. I am indebted to Matt Becker, the press's editor-in-chief, for his constant support, guidance, and encouragement as my rough unsolicited draft was transformed into a finished product. I also thank Sally Nichols, the designer who dealt with the book's many figures; Rachael DeShano, who steered the project from approval to printing; Chelsey Harris, who aided with marketing and publicity; and Dawn Potter, the copy editor, who had to wade through and correct my many typos and grammatical errors and whose considerable efforts have made this a much more readable text. In addition, I am grateful to the anonymous reviewers who offered valuable feedback; their suggestions for revisions and additions made a significant difference in the final manuscript.

This book was inspired by many trips to New England and Acadia National Park with my wife and fellow forest enthusiast, Chris. I thank her for being both a hiking companion and a patient sounding board as this book went through endless rewrites and reconfigurations. Even when I thought this project would never end, she was confident that I was wrong. As usual, she was right.

Conifers of the New England–Acadian Forest

Introduction

The New England–Acadian Forest Ecoregion

THE YEAR WAS 1853, and Henry David Thoreau had returned to the woods of Maine.

> It rained all this day and till the middle of the next fore-
> noon, concealing the landscape almost entirely; but we had
> hardly got out of the streets of Bangor before I began to
> be exhilarated by the sight of the wild fir and spruce-tops,
> and those of other primitive evergreens, peering through
> the mist in the horizon. It was like the sight and odor of
> cake to a schoolboy.[1]

Exhilarating indeed! Thoreau was at the start of a second journey into an ecological domain known today as the New England–
Acadian Forest ecoregion, an ecosystem filled with spruce, fir,
pine, cedar, hemlock, and northern hardwood species. A few
loggers had moved into the region north of Bangor to harvest the
centuries-old white pines, but Thoreau could still find extensive
mature forests with a tree species composition almost unchanged
for a thousand years or more. Lucky Henry David. Today, little
of the old growth is left, but the Acadian Forest ecoregion still
exists in the form of diverse, second-growth woodlands filled
with younger versions of the same tree species as in Thoreau's
day, and these can be found and appreciated in places such as
Acadia National Park.

This ecoregion takes its name from the land known to seven-
teenth- and eighteenth-century French colonists as Acadie. When
the territory was controlled by the French Empire, it included all

1

of today's Nova Scotia and, by some accounts, extended into parts of present-day New Brunswick and eastern Maine. In the 1700s, following a series of wars between the French and the English, control of Acadie, or Acadia, passed to the British Empire, but the name remained tied to the area and its forests, which today extend over a much larger area than the French colonial territory did. In Canada, the forest covers most of Nova Scotia, Prince Edward Island, and New Brunswick as well as a slice of eastern Quebec. South of the border, the ecoregion stretches from Maine through New Hampshire to Vermont and down into western Massachusetts. Because the word *Acadia* retains a strong link to Canada, many ecologists refer to the ecoregion as the New England–Acadian Forest (NEAF) ecoregion, especially when discussing forests located in the United States.[2]

The NEAF is an ecological transition zone sandwiched between the cold-climate boreal forests to the north and the temperate hardwood forests to the south. As a result, it includes a combination of northern and southern tree species. Among them are northern hardwoods such as quaking aspen, paper birch, sugar maple, and beech as well as several types of conifers—the subject of this book. Red spruce (*Picea rubens*) is a key species in this ecoregion; if you see a lot of red spruce in a northeastern forest, you'll know that you're probably standing in the New England–Acadian Forest. Other common conifers include the aromatic balsam fir (*Abies balsamea*), the towering eastern white pine (*Pinus strobus*), the shade-tolerant eastern hemlock (*Tsuga canadensis*), the rot-resistant northern white cedar (*Thuja occidentalis*), and the sandy soil–adapted red pine (*Pinus resinosa*). As is typical of ecological transition zones, the Acadian Forest also contains conifer species that are near the edges of their geographic distributions. For instance, pitch pine (*Pinus rigida*) is a mid-Atlantic species and is more characteristic of the Northeastern Coastal Forest ecoregion to the south. However, it can also be found in the southern margins of the Acadian Forest, especially on rocky sites along the Maine coast, where it is near the northern limit of its range. White spruce (*Picea glauca*), black spruce (*Picea mariana*), jack pine (*Pinus banksiana*), and tamarack (*Larix laricina*) dominate the northern, or boreal, forests of Canada, and these conifers are close to the southern edges of their continuous ranges in northern New England and New York. When growing along the Maine coast, they probably benefit from the effects of the cold water of the Gulf of Maine, which keeps the coast cool and damp in the summer.

Humans are a part of ecosystems, too. We interact with the other species in a given region and incorporate many of them into our lives. This is certainly true in the Acadian Forest. Due to their geographic distribution and their physical and chemical properties, the region's conifers have played significant roles in the culture and history of New England, southeastern Canada, northern Europe, and beyond. As this book will show, these trees have tales to tell, and their stories will take us many centuries into the past and across a wide geographic area. We'll begin in the 1530s, when French explorers overwintering in the Saint Lawrence River valley were dying of scurvy until a local Iroquois man showed them how to make a vitamin C–rich tea from the foliage of a conifer they came to call the "tree of life." We'll look at how New England's tall white pines became a critical resource for the Royal Navy, which used the trees for masts as large as four feet in diameter and more than one hundred feet in height. But local colonists had other ideas about how to use the pines, and their disagreements sparked conflicts that lay behind the American Revolution. After the colonies gained independence, white pines continued to influence Anglo-American relations. The 1783 Treaty of Paris was vague about the location of the boundary between Maine and the British colonies to the north and east, and this created a 12,000-square-mile swath of contested land. Though few people lived in the area, the value of the land and its pines spurred the young state of Maine to take action in 1839, leading to a conflict known as the Aroostook War. White pines would fuel the development of Maine's timber industry in the first half of the 1800s, stimulate the growth of the boom town of Bangor, and prompt timber barons to build dams and canals to drive logs down to mills along the Penobscot River. Later in the century, as white pine supplies began to diminish, loggers turned to red spruce for lumber and for papermaking pulp, developing new harvesting technologies such as steam-powered log haulers with an innovative tread system.

Conifers are not merely valuable for their wood products; they also produce aromatic chemicals called terpenes. The trees construct terpene molecules to protect themselves from herbivores and disease-causing microbes, and it is these compounds that give spruce beer its piney flavor, spruce gum its chewiness, and balsam fir the wonderful aroma that made the species the foundation of the Christmas tree industry in northern New England. As this book will show, antimicrobial properties may also be present in the resin extracted from balsam fir trunks; certainly it was used medicinally for centuries before humans learned that bacteria and other microorganisms could cause disease. The resin, known as Canada balsam, played a role in the development of germ theory, which revealed a link between microbes and illness. Microscopists also used Canada balsam to glue together glass lenses and thus create permanent microscope slides for study. These developments inspired a young surgeon who would eventually revolutionize surgery by using carbolic acid to kill microbes before they could cause deadly surgical infections.

Pitch pines are another source of terpene-rich resins. These resins can be converted into tar, pitch, turpentine, and rosin, products known collectively as *naval stores* because they were essential to the function of wooden sailing ships. As the British Empire became increasingly dependent on its American colonies as a source for these commodities, their price per barrel rose, and colonists took to extracting turpentine from pitch pines in a murky border region between Connecticut and Massachusetts. This led to brawls and arrests as residents of each colony accused their border neighbors of trespass and illegal turpentine collection. The dispute forced representatives from each side to agree to draw definitive map boundaries. Pitch pines continued to have political ramifications later in the century: when colonists tarred and feathered their political opponents, they used pine tar as the adhesive.

The eastern hemlock is an important source of a different group of valuable plant compounds: the tannins. Tannins protect trees by binding themselves to proteins, a reaction that inhibits herbivores' digestion of leaf material. You can test this phenomenon for yourself by drinking a glass of red wine. As the grape-skin tannins bind to proteins in your mouth, they produce a drying and puckering effect. Tannins also bind themselves to the proteins in animal hides and skins. Thus, for tanners during the nineteenth century, the hemlock forests of eastern Maine were the foundation of a leathermaking boom.

The conifers described in this book are found throughout the New England–Acadian Forest ecoregion. However, a visit to the diverse forest habitats of Maine's Acadia National Park is a way to see all of them in a particularly beautiful setting. George Dorr, the park's first superintendent, promoted it as a place where the Acadian Forest "is typically represented with singular completeness." In his book *The Acadian Forest*, he described Mount Desert Island, where most of the park land is located, as a setting "where land and sea conditions meet and where a unique topography creates a correspondingly exceptional range of woodland opportunity." From the start, Dorr felt that "a constant aim" of Acadia National Park should be "to establish on the Island . . . a permanent exhibit of this forest growing under original conditions."

> Such an exhibit has extraordinary value. A forest is far more than the mere assemblage of its trees; associated with them it contains, in regions of abundant moisture such as the Acadian, a related life, both plant and animal, of infinite variety and richness, whose home and sheltering habit it makes. . . . Such a forest is a wonderful complex of mutually dependent forms, a complex anciently

established which once obliterated in a region can never be restored.

The typical trees of the Acadian forest, those that give it its peculiar character, are the northern evergreens, the cone-bearing pines and firs and spruces, the hemlocks and the arbor vitae [northern white cedar]. It is of these [that] one thinks in picturing to oneself the region. . . . Longfellow set the Acadian scene for us in *Evangeline* with "This is the forest primeval, the murmuring pines and the hemlocks," and far out to sea in early long-voyaged days the approaching sailor welcomed with delight the pungent forest fragrance.[3]

As Dorr noted, the location and topography of Acadia National Park create the ideal location for seeing the full range of Acadian Forest conifers, including a few conifer species not covered in detail in this book. Therefore, this book concludes with an appendix describing where to find these trees in the park, along with a guide to their identification.

Chapter 1

A Tree of Life for
French Explorers

I T WAS DECEMBER 1535. A few more than a hundred
Frenchmen were huddled along the Saint Lawrence River near
present-day Quebec City—some spread among three ice-bound
ships, others inside a small fort beside the river. Their leader, the
French-Breton explorer Jacques Cartier, had recently learned that
a "pestilence" had broken out among the Laurentian Iroquois in
the nearby village of Stadacona. It's likely that the news made
Cartier uneasy. The French were a long way from home, their
ships were locked in ice, and they were in no position to deal
with a dangerous and spreading illness. What Cartier didn't know
was that they were also about to experience the coldest, harshest
winter of their lives, an experience that must have shocked them,
given that Stadacona's latitude was farther south than Paris's.

Cartier, the first European to map the Saint Lawrence, was
part of France's effort to join the rest of Europe in exploring,
colonizing, and exploiting the lands of the western hemisphere.
Under the aegis of King François I, he and other French explorers hoped to find gold, discover a northwest passage to Asia, or
establish trading networks for North American furs and other
commodities. Already, in 1534, Cartier had sailed into the Gulf of
Saint Lawrence but had retreated to France before the end of the
year. On his return to the region in the spring of 1535, he explored
upstream as far as modern-day Montreal, and by October he and

his men had settled in near Stadacona. Their first winter in North America would be a nightmare.

The trouble began with the December illnesses in Stadacona. After hearing that as many as fifty Iroquois had died that month, Cartier acted to protect his party, forbidding contact between the French and the tribe. This proved to be a futile gesture as the malady afflicting the Stadacona people was probably scurvy, a potentially fatal disease caused by a lack or deficiency of vitamin C (ascorbic acid). Cartier tried to act for the best, but there was nothing to be gained by prohibiting contact: scurvy isn't contagious or caused by pathogenic microbes. While it's possible that some of the deaths in Stadacona were due to infectious diseases, they likely came from the French themselves so were already present in the ships and the fort. (For the next several centuries, European diseases would have a devastating effect on the native peoples of North America.)

It's hard to blame Cartier for trying something, anything, to save his men, but breaking off contact with the Iroquois did not protect the shivering French. Scurvy soon devastated them. The disease leads to blood disorders and an inability to produce the collagen that literally holds our bodies together, and Cartier recorded a near-textbook description of its ravages: "Some lost all their strength, their legs became swollen and inflamed, while the sinews contracted and turned black as coal. . . . The legs were found blotched with purple-colored blood. . . . And all had their mouths so tainted that the gums rotted away down to the roots of the teeth which nearly fell out." In mid-February 1536, Cartier wrote, "Of the 110 men forming our company, there were not ten in good health so that no one could aid the other, which was a grievous sight considering the place where we were. For the people of the country [the Iroquois] who used to come daily to the fort saw few of us about." It's clear that Cartier was worried about more than just scurvy. He knew that the villagers recognized that the French were in bad shape and vulnerable to attack:

"Not only were eight men dead already, but there were more than fifty whose case seemed hopeless." Survivors were too weak to bury the dead in the frozen ground; instead, they scratched out holes in the snow for the bodies.[1]

In desperation, Cartier ordered autopsies of the scurvy victims "to see if anything could be found out about it." Examinations revealed numerous effects of vitamin deficiency on internal organs, including a heart that "was completely white and shriveled up." But these postmortems offered no answers or cures. By late winter, twenty-five men had died—nearly a quarter of the entire group—and Cartier had "little hope of saving more than forty others." The French "had almost lost hope of ever returning to France." Nonetheless, Cartier tried to maintain a show of strength. Throughout the ordeal, Cartier and a handful of others had regularly paraded outside of the fort, working to create the illusion that the French had not weakened.[2]

One day, as Cartier was making his usual circuit across the ice and snow, he saw a man approaching him from Stadacona. This man, Dom Agaya (or Domagaya), was no stranger; Cartier had met him in 1534, during his first voyage to North America. The French had been exploring along the Gaspé Peninsula at the mouth of the Saint Lawrence when they encountered a group of Laurentian Iroquois, led by the sachem Donnacona, who had come downriver from Stadacona on a fishing expedition. After doing some trading with the Iroquois, Cartier made a diplomatic misstep: he raised a thirty-foot-tall cross decorated with the French fleur-de-lys, a coat of arms, and the words "Vive le Roi de France." The Iroquoians were not pleased by the suggestion that the king of France had possessive feelings about their land. Nor were they pleased when the French ended the encounter by seizing two Iroquois men and transporting them to France. Perhaps Cartier had convinced himself that the men were being shipped to Europe as guests, but the event was essentially a kidnapping. In France, the two abducted men had learned a little French; and when Cartier returned to the Saint Lawrence in the summer of 1535, he brought them back to their homeland. They could now act as interpreters and mediators with the Stadacona villagers.

One of those men was Dom Agaya, and Cartier wrote of being "delighted" to see him looking well. In fact, he must have been amazed because, just ten or twelve days earlier, Dom Agaya had been "extremely ill with the very disease [Cartier's] own men were suffering from." His gums "had rotted and become tainted," his sinews had contracted, and "one of his legs about the knee had swollen to the size of a two-year-old baby." This was a curious unit of measure, but, regardless, here he was, walking across the March snow, apparently the picture of health. What had saved him?[3]

Dom Agaya explained that "he had been healed by the juice of the leaves of a tree and dregs of these, and that this was the only way to cure sickness." Immediately, Cartier asked for a sample of these materials, claiming he needed them to "heal his servant who had caught the disease when staying in Chief Donnacona's house [in Stadacona]." This was a ruse, as no one had stayed in the chief's house and almost the entire French contingent was suffering from scurvy. However, Dom Agaya agreed. He sent two village women and a French captain to gather nine or ten branches from a certain conifer, and Cartier's party was shown "how to grind the bark and the leaves and to boil the whole in water." The afflicted men were to "drink [the boiled extract] every two days, and place the dregs on the legs where they were swollen and affected." According to Cartier, Dom Agaya called this tree, "in their language, *Annedda*," and claimed that it "cured every kind of disease."[4]

What was the effect of the annedda tea on the scurvy-ridden French? Initially, the sailors resisted drinking the concoction. But their reluctance vanished after a few tried it. Cartier wrote that they "recovered [their] health and strength and were cured of all the diseases they had ever had. And some of the sailors who had been suffering for five or six years from the French pox [syphilis] were by this medicine cured completely." As word of the drink's effect spread, "there was such a press for the medicine that [the men] almost killed each other to have it first. . . . It benefited us so much that all who were willing to use it recovered health and strength, thanks be to God." Cartier reported that the drink "produced such a result that had all the doctors of Louvain and Montpellier been there, with all the drugs of Alexandria, they could not have done so much in a year as did this tree in eight days."[5]

Certainly, the tea did not cure syphilis, a disease that was killing a substantial number of Europeans in the sixteenth century, but Cartier's exaggeration was a reflection of his genuine excitement at the improvement in his men's condition. Though he did not understand this at the time, the tea contained vitamin C, a potent cure for scurvy. Within a few weeks, it saved dozens of men who had suffered for months and were near death. To the French, this was nothing short of a miracle. The annedda tree was truly a "tree of life"; and it became known, in Latin, as *arbor vitae* and, in French, as *l'arbre de vie*.

It seems surprising that a man who had been kidnapped by the French would step forward to save them. There is no record of Dom Agaya's thoughts, so we can only speculate about what led him to help Cartier. He and the tribe may have acted from compassion or have calculated that the French could prove useful. Earlier contacts with Europeans had proven that the strangers were both a threat and an opportunity. Already, for decades, fishing fleets from Europe had been trading with tribes along the Atlantic coast of North America, with the Natives offering furs and hides in exchange for glass beads, knives, and axes. If the Saint Lawrence Iroquois could establish a trading alliance with the French, the resulting economic benefits—including sharp-edged weapons—might improve the group's chances of surviving an intertribal war. Their rivals were certain to acquire such tools, and the villagers of Stadacona would be wise to do the same.[6]

Whatever the motive, Dom Agaya had used his knowledge about the tree of life to save Cartier's crew; and if the tale had featured a Hollywood ending, he would have been generously rewarded for his efforts. Unfortunately, however, Cartier did not exhibit much gratitude. In May 1536, as the eighty or so survivors prepared to sail back to Europe, the French kidnapped Donnacona and three others from Stadacona, including Dom Agaya, and carried them away to France. Cartier may have intended to bring them back in 1537, but wars and other events intervened, and it would be five years before he returned to the Saint Lawrence. By then, most of the transported Iroquois, including Dom Agaya, had died, probably from European diseases, to which they had little resistance. None ever returned to Canada.

In 1541, Cartier returned to the Saint Lawrence River on his third and final voyage, this time serving as chief navigator in an expedition under the command of Jean-François de la Rocque de Roberval. In May, he forayed ahead of the rest of the expedition, traveling to the Stadacona area, where he built a fort and established a colony he named Charlesbourg-Royal (at the present-day site of Cap-Rouge in Quebec City). On this visit, the local Iroquois were far less cooperative. Cartier's habit of kidnapping their people and his tendency to settle on Iroquois land without permission had significantly increased their hostility. Moreover, they had no need to interact with the colonists at Charlesbourg-Royal; by this time, desirable European goods could be acquired via trade at eastern points of exchange. In any event, Cartier didn't show much interest in trading for furs. His goal was to find gold and other minerals, both for his own enrichment and because the French royalty was searching for territory that could match the staggering output of the Spanish mines in Central and South America. Yet Cartier already knew the secret of the annedda tree. There is no surviving written account of events in the colony during the winter of 1541–42, but other writings about the third voyage confirm that he remembered the tree of life. And as far as we know, the colonists did not suffer from an outbreak of scurvy.

By early 1542, Cartier and his men had collected large amounts of gold and diamonds. Or that's what they thought. Eventually, they learned that the supposed gold was just pyrite (iron sulfide, or fool's gold) and the diamonds were common quartz crystals—a discovery memorialized in the French expression *faux comme les diamants du Canada* (as false as diamonds from Canada). In the meantime, however, Cartier believed he possessed a treasure, and he was ready to give up exploring and cash in. That spring, he sailed from North America for the last time. On his way home, he crossed paths with the Sieur de Roberval and his ships along the Newfoundland coast. Roberval commanded Cartier to join him in going back up the Saint Lawrence to Charlesbourg-Royal. Cartier chose to decline this "opportunity." In Roberval's words, "he and his company, moved as it seems with ambition, because they would have all the glory of the discovery of those parts themselves, stole privily [secretly] away the next night from us, and without taking their leaves, departed home for Bretagne [Brittany]."[7] Roberval continued on to the Charlesbourg-Royal fort, which he renamed France-Roy. Then, during the winter of 1542–43, he recorded that "many of our people fell sick of a certain disease in the legs, loins, and stomach, so that they seemed to be deprived of all their limbs and there died thereof about fifty." It's difficult to be certain from this brief description, but it seems likely that the French community, numbering about two hundred sailors, soldiers, and settlers, lost nearly a quarter of their people to scurvy that winter.

Sneaking away from Roberval's fleet had been bad enough. Even worse, it appears that Cartier had failed to pass along mission-critical information about the value of the annedda tree. Not surprisingly, both the fort and the colony at France-Roy were abandoned in the late spring of 1543, and Roberval returned to France. Then, about seventy years later, another group of Frenchmen would again experience the terrors of scurvy. Tragically, critical details of what Cartier had learned had been largely forgotten.

After the failures of Cartier and Roberval, the French made few attempts to establish permanent settlements in northern

North America in the remaining years of the sixteenth century. Then, in the early 1600s, a group organized and led by Pierre Dugua (or Du Gua), Sieur de Monts (or de Mons), returned to the region to sow a few colonies. De Monts also hoped to earn money by means of a royally granted monopoly on the North American fur trade. In May 1604, his group reached the coast of La Cadie, or l'Acadie (in English, Acadia). At the time, this territory included parts of present-day Nova Scotia, New Brunswick, Maine, and the Bay of Fundy. After exploring the bay and the Nova Scotia coast, they established a base on a five-acre island dubbed Saint Croix, located near the mouth of a waterway later called the Saint Croix River, which now forms a section of the border between Maine and New Brunswick.

De Monts chose Saint Croix Island as the location for a fort and a tiny settlement because it could be defended against Spanish or English incursion and was geographically well positioned to become a trading post. In September 1604, as he led efforts to raise buildings and establish gardens on the island, de Monts dispatched a small ship called a patache to explore the

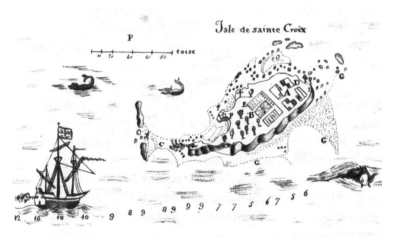

FIGURE 1. Samuel de Champlain's map of Saint Croix Island, 1613. Its cemetery, filled with scurvy victims, is marked with a cross and the letter *E*. In the collection of the Library of Congress.

coast of present-day Maine. With two masts and a shallow draft, patches were light and quick, perfect for coastal investigation, and this one was under the command of a capable navigator, cartographer, soldier, and chronicler named Samuel de Champlain.[8]

As Champlain sailed south, he caught sight of an island: "It is very high and cleft in places, giving it the appearance from the sea of seven or eight mountains one alongside the other. The tops of most of them are bare of trees, because there is nothing there but rocks. The woods consist only of pines, firs, and birches. I named it Mount Desert Island. Its latitude is 44O 30."[9] Today this is the home of Acadia National Park. Champlain was not the first European to see the island, but his name for it is the one that has carried down to the present day. In the original French, Mount Desert Island is *Ile des Monts Déserts,* and the word *déserts* refers to its barren, or desert-like, mountaintops, suggesting that their near-treeless condition dates back to at least the early 1600s.

Shortly after spying the island, Champlain nearly missed his chance to leave his mark on the maps of Maine. As his vessel approached the southeastern corner of Mount Desert, a grinding noise alerted the crew to the fact that the patache had run into something hard, just below the surface of the water. It's likely that the ship had hit the Spindle, a rocky shelf that today appears at low tide off Otter Point. Apparently, high tide had hidden the rocks from the sailors, and perhaps the crew had also been distracted by the beauty of the mountains. If the patache had gone down with all aboard, the name *Isle des Monts Deserts* might have died with the ship. But the vessel stayed afloat long enough to round the point, allowing the crew to ground it on the flats in Otter Cove. When the tide went out, the vessel gently settled on its side, allowing the French to inspect and repair the hole in the hull, and they sailed out of the cove the next day. After a little more coastal exploring, Champlain rejoined Sieur de Monts's company on Saint Croix Island in October 1604.

The island may have been a defensive haven, but unfortunately it offered no protection from scurvy, a foe as deadly as the

English or Spanish. That winter, the French found themselves trapped on a dot of land in the middle of a frozen river, subsisting primarily on a diet of salt meat, which contained little or no vitamin C. They had had hoped to avoid this fate. In early summer, they had sowed vegetable seeds that could have supplied the essential vitamin. But as Champlain noted, "the island was nothing but sand, [and] everything was almost burnt when the sun shone."[10] It's not clear how much garden produce the French were able harvest, given the short growing season. It's also possible that they had a supply of European fruit products that contained vitamin C: archaeological excavations on the island have uncovered storage containers that may have held preserved fruit. But whatever they had was not enough.

In his journal, Champlain noted that the men soon began to suffer from "a certain malady . . . called land-sickness, otherwise scurvy, according to what I have since heard stated by learned men." He recorded the classic symptoms of the disease: putrefied gums, teeth that "could be drawn out with the fingers without causing pain," weak and contracted leg muscles, abdominal pains, and shortness of breath. Eventually, the men "were in such a state that the majority of the sick could neither get up nor move, nor could they even be held upright without fainting away." And then they died.

The French on Saint Croix were reliving Cartier's nightmare; and, apparently, autopsy was the go-to move when members of a European expedition died of scurvy during a North America winter because the bodies of some of the dead were cut open and inspected. These postmortem examinations showed extensive internal damage, but, as Champlain noted, "we could find no remedy with which to cure these maladies," and the surgeons who performed the dissections "were unable to treat themselves so as not to suffer the same fate as the others." Four hundred years later, archaeologists would find evidence that the French had searched everywhere in the bodies for answers when researchers uncovered a skull with the top half cut through and then replaced before burial. By spring, when new plants finally gave the men a change in diet, thirty-five of the original seventy-nine had died. Within months, the tiny island had become the home of an enormous cemetery.[11]

The expedition under the Sieur de Monts lost 45 percent of its men, mostly from scurvy, a death rate higher than that in Cartier's company. But of course Cartier had learned the miracle of the annedda tree. So why didn't the French at Saint Croix have access to this miraculous cure? Perhaps the island itself was too small to support a population of the tree-of-life conifers; perhaps all the trees on the island had been felled for building materials. Almost certainly, however, the trees would have grown along the riverbanks, though the French would have struggled to reach them across the ice.

We do know that Champlain had read about Cartier's experiences. He was aware of the scurvy outbreak of 1535–36, and he had learned about the annedda tea. All he had to do was find the tree, but apparently he did not know *which* tree was the correct one. How could such vital data have been lost? We have no clear answer, but several factors may have contributed to this disastrous situation.

Significantly, Cartier's surviving writings contain no detailed description of the annedda tree. Either he never recorded that information or, more likely, the record was lost so wasn't available to early seventeenth-century explorers. Still, we have evidence that at least a few people in France knew about this tree before 1605: for instance, in 1553, a physician named Pierre Belon wrote about a Canadian-sourced *l'arbre de vie* that was growing in the royal garden at Fontainebleau.[12] Given the miracle cure at Stadacona (not to mention the tree's alleged ability to cure syphilis), it seems likely that Cartier had brought home samples of the tree. But tragically, that chain of information had broken by the time of the Saint Croix outbreak.

Moreover, by 1600, the Laurentian Iroquois were no longer living in the Saint Lawrence River valley. It's not clear what happened to them. They may have been driven out by war with another tribe or been decimated by European diseases. Regardless, they were gone. When Champlain visited the site of Stadacona in 1603, there was no one left to share the secret of the annedda tree. During his explorations along the coast of Maine, he had asked about the miracle plant, but the Algonquin-language tribes in the region didn't know of a tree linked to the word *annedda*. Marc Lescarbot, another chronicler of the winter at Saint Croix, wrote: "As for the tree called *Annedda*, of which the said Cartier speaks, the [inhabitants] of these regions know it not."[13]

In pondering the deaths on Saint Croix, Champlain concluded that a diet heavy in salted meat had played a part. He was on the right track: yes, a diet composed of nothing but salted meat will lead to scurvy. But he went wrong when he presumed that the problem was due to poisons in the meat that had somehow corrupted the blood. It was many years before people learned that scurvy is a deficiency disease, not a poisoning disease.

After the terrible winter of 1604–5, the French at Saint Croix moved their base of operations across the Bay of Fundy to the northwestern coast of today's Nova Scotia, naming their new home Port Royal. Because they were now on the mainland, they were able to vary their diet during the winter of 1605–6 by exchanging goods with the Mi'kmaq groups that dominated the peninsula. Since Cartier's day, trade between the French and the tribes in the Acadian Forest region had expanded significantly and relatively peacefully. Europeans had developed an insatiable desire for broad-brimmed felt hats, which had led to a surge in demand for North American beaver pelts because these contained the downy fur ideal for making hat felt. In exchange for these pelts, traders now offered copper kettles, which were as popular among the coastal tribes as felt hats were on the other side of the ocean. The pelts and other North American products were collected and transported to Europe by whalers and cod fisher-men, who were diversifying their business via the furs-and-kettles exchange. It probably helped that the sailors came ashore only very briefly, just long enough to salt fish, process whale blubber into oil, and do a little trading. They could exhibit good manners with their trading partners because they were uninterested in conquering or holding territory.

At Port Royal, the French appear to have benefited from decades of trade; their relations with the Mi'kmaq were better than those between Cartier and the Iroquois. A sagamos (or sagamore) was the leader of the local Mi'kmaq group. Known to the colonists as Henri Membertou, he often dined with the French in their Port Royal compound. Marc Lescarbot recorded that, "as for the Sagamos Membertou, and the other chiefs, who came from time to time, they sat at table, eating and drinking like ourselves. And we were glad to see them, while, on the contrary, their absence saddened us."[14] We do not know if the Mi'kmaq shared foods that were high in vitamin C, as Champlain still had muddled and mostly erroneous ideas about diet and thus did not record that information. However, during the first winter at Port Royal, he reported twelve deaths, with autopsies confirming that most were due to scurvy. This was a much better outcome than the previous winter's. Nonetheless, between 20 and 25 percent of the company were buried.

In 1606, the French benefited from a full growing season at Port Royal and were able to harvest crops and gather and preserve fruit. This seems to have improved their diet. Lescarbot noted, "For our rations we had peas, beans, rice, prunes, raisins, dried cod, and salt meat, besides oil and butter." These fruits and vegetables would have provided some vitamin C, although cabbage would have been a better source. Possibly it, too, was grown at Port Royal as it was a staple in seventeenth-century France, but we do not have direct evidence from Lescarbot's list. We do know he believed that the men's health was compromised by another lack at Port Royal:

> One further preservative is necessary to complete a man's
> content and to fill up his pleasure in his daily talk, which is
> that each man should have the honorable company of his
> married wife; for if that be lacking, the good cheer is not
> complete, one's thoughts turn ever to the object of one's

love and desire, home-sickness arises, the body becomes
full of ill-humors, and disease makes its entrance.[15]

But wives were not available so, during the winter of 1606–7,
Champlain acted on his own assumptions about scurvy. Hoping
to reduce the dependence on salted meat, he worked to create
a steady supply of fresh meat at Port Royal. To encourage this,
he established the *Ordre de Bon Temps* (Order of Good Times),
featuring "a chain which we used to place with certain little cere-
monies about the neck of one of our people, commissioning him
for that day to go hunting."[16] Lescarbot recorded the ceremony
that took place after a successful hunt: "The ruler of the feast or
chief butler . . . marched in, napkin on shoulder, wand of office
in hand, and around his neck the collar of the Order, which was
worth more than four crowns; after him all the members of the
Order, carrying each a dish."[17] At the end of the day, the collar
was passed to the next man in line, who took the lead in providing
the main meal for the following day.

The *Ordre de Bon Temps* seems to have been an effective
motivation. Champlain observed, "All vied with each other to
see who could do the best, and bring back the finest game."[18]
Lescarbot reported that the men hunted a wide variety of bird
and mammal species, "whereof we made dishes well worth those
of the cook-shop in the *Rue aux Ours* [Street of Cooked Meat
Sellers]."[19] Champlain thought that fresh meat and fish were
good medicine, and in fact the muscle tissue of ducks and other
waterfowl does contain some vitamin C. Although the flesh of
the region's mammals contains almost none, liver, kidneys, and
other organ meats do. Whether or not the men absorbed any of
this available vitamin C is an open question, given that ascorbic
acid is heat sensitive and can be destroyed by cooking. But perhaps they ate fresh liver and preserved fruits along with their
steaks, for Champlain reported that deaths from *mal de la terre*
dropped significantly that winter—down to seven, as confirmed
by the now obligatory autopsies. The *Ordre de Bon Temps* had

improved the settlers' diet and had provided good cheer as well, which Lescarbot saw as a preservative of health. But Champlain never did find the annedda tree.

So, which conifer species was it? This question has challenged botanists and historians for years. Even today, no one is sure. The identification problems began with Cartier. We know that scurvy repeatedly devasted French parties overwintering in northern North America. Thus, detailed descriptions of the scurvy-curing tree should have been a priority for those exploring and colonizing the region. Yet, as noted, any recorded account has been lost. Cartier was no botanist, and he probably would not have been able to identify and describe trees to the species level. However, he did know how to distinguish among different conifer genera. His writings include many passages about Canadian flora in the Saint Lawrence River valley, demonstrating that he was familiar with conifer types such as cedars, pines, and spruces. Though he likely grouped the balsam firs and hemlocks with the spruces, he recognized the difference between what he called a "spruce" and a cedar. Yet he does not use any of these terms when referring to the annedda tree. His most specific description appears in a passage from his account of his third voyage: "On both sides of the said river . . . there is one kind of tree above three fathoms [in circumference], which they in the country call *Anneda*, which hath the most excellent virtue of all the trees of the world, whereof I will make mention hereafter."[20] But there is no "hereafter" in the known writings of Jacques Cartier, which may explain why de Monts and Champlain were left up the Saint Croix River without a paddle.

So what have contemporary botanists concluded about the identity of the tree? A leading contender is the northern white cedar (*Thuja occidentalis*), listed in the appendix. This species grows throughout the southern part of eastern Canada and across the northern tier of the United States, from Wisconsin to Maine. Because the seedlings grow well in cool, moist soils and on rotting wood, the trees are often found along streams, rivers, and lakes

and in the moist soil of upland forests. White cedar can tolerate soils that may be too wet for other trees, which explains its
nickname: swamp cedar. Yet it can also adapt to rocky summits
where soil is thin, dry, and low in nutrients. Because white cedar
can survive under trying circumstances, it sometimes becomes
a dominant part of the plant communities on cliffs, especially
if the bedrock has seeps or water-collecting depressions. In the
area around present-day Quebec City, there is abundant riverbank habitat as well as many bluffs and cliffs. This suggests that
northern white cedar would have been readily available to the
Iroquois in Stadacona.

As mentioned, we know that there was a tree of Canadian origin growing at Fontainebleau, which Belon identified as *arbor vitae*
or *l'arbre de vie*. It is easy to posit a link between Cartier's annedda
and the this Canadian transplant. Belon's writings describe the
tree as carrying *foliis applanatus*, or flattened leaves, a description
that is consistent with northern white cedar. Not only does the
name *tree of life* suggests significant healing powers, but the word
arborvitae is still linked to the northern white cedar, often used
as an alternative common name.

However, botanists and historians have also considered white
pine, red pine, white spruce, black spruce, hemlock, and balsam
fir as the source of Cartier's life-saving tea. The needles of many
northern forest conifer species contain ample amounts of vitamin
C, and accounts from other times and tribes describe treating
scurvy with teas made from some of these conifers.[21] In fact, if
Champlain had made tea from the needles of almost any local
conifer, he might have saved dozens of lives. Perhaps he feared
poisoning or was stymied by the difficulty of traveling across the
ice but, under the circumstances, the French had little to lose by
playing conifer roulette.

White cedar foliage has about 50 milligrams of ascorbic acid
per 100 grams of leaf material, but red spruce (*Picea rubens*),
hemlock (*Tsuga canadensis*), and balsam fir (*Abies balsamea*) all
have much higher levels, with balsam fir topping the list: it has

roughly 260 milligrams per 100 grams of needles. This line of evidence has persuaded the historian Jacques Mathieu that *A. balsamea* was the annedda tree.[22] The very high vitamin concentration in fir tea might explain why Cartier's men recovered from scurvy so rapidly.

Long before the science of chemistry emerged, people around the world were working out plant-based treatments for disease via empirical observation or trial and error. Perhaps such methods led northern groups of Native Americans to conclude that balsam fir offered the best treatment for scurvy. Still, we cannot eliminate red or black spruce as annedda candidates because both species were used in the 1700s and 1800s to brew a beer used to treat scurvy. In the end, we'll probably never have a definitive answer to the annedda mystery.

Chapter 2

White Pines and the King's Broad Arrow

The [White] Pine has been appropriately called the Monarch of the Forest. Taken all in all, it is the crowning master-piece of all woody plants.... I was reared among the noble Pines of Maine, nestled in my cradle beneath their giant forms, and often has the sighing wind made music that has calmed me to repose as it gently played through their tasseled boughs. Often have I been filled with awe as I gazed upon their massive trunks and raised my eye to their cloud-swept tops.... In strength the Oak excels, but in towering grandeur and massive diameter the Pine far exceeds the Oak, and indeed all other North American trees.

—John Springer, *Forest Life and Forest Trees*

WHEN THE ENGLISH FIRST stepped ashore in North America, white pine (*Pinus strobus*) was the king of the New England forests (see figure in the appendix).[1] Old-growth pines could be up to five feet in diameter and between 150 and 160 feet tall—the largest conifers in the Northeast. The sight of such trees triggered amazement and delight. But in the 1600s, those long straight trunks also triggered another thought: ships' masts. This was not a poetic vision; but if your nation's defense depended on a navy composed of wooden ships, you might be forgiven for your utilitarian view of these natural wonders.

England lived and died by the strength of the Royal Navy, and with every passing century its warships grew larger; by the mid-1600s, the largest carried mainmasts that were up to forty yards in height. A rule of thumb stated that a mast should be an inch in diameter for every one yard of height, so a forty-yard mast needed to be forty inches in diameter at its base. But wooden ships were vulnerable in storms and battle, and the steady loss of masts meant that the Royal Navy needed many of them.

While the hull and decks of sailing ships were often constructed from hardwood timber such as oak, the masts were almost always hewn from the more flexible trunks of conifers. Centuries of logging had nearly eliminated huge conifers from the British Isles, so the Royal Navy had to find these trees in other countries. During the first half of the 1600s, the English were heavily dependent on conifer trunks imported from forests around the Baltic Sea and the west coast of present-day Norway (then part of the Swedish Empire). The conifers from these regions were broadly known as Baltic firs, but most were probably Scots, or Scotch, pine (*Pinus sylvestris*). Unfortunately, the imported trunks were often under thirty-six inches in diameter, and thus a single tree was too skinny to make a forty-yard-high mast. Large-diameter "made masts" had to be constructed from multiple small-diameter trunks that were cut, fitted together, and wrapped with iron bands. These composites worked, but shipbuilders preferred to make a mast from a single tree.[2,3]

Moreover, a heavy reliance on trees from the Baltic region made English naval officials anxious. What if their supply lines were cut? The Royal Navy's vulnerability became clear during the First Dutch War (1652–54), when Denmark, allied with the United Provinces of the Netherlands, blocked the narrow passage between the Baltic and North seas. This greatly reduced the importation of trees vital to the English war machine, a potentially disastrous situation. England survived the war, but even in peacetime most of the pine forests remained in the hands of the regional powers, Sweden and Russia, who were not always friendly to Britain. So the English began to search for other sources, and their eyes turned to the old-growth white pines of New England.

Why were white pines so abundant in northeastern North America? *Pinus strobus* is well adapted to cool climates such as New England's and can grow on diverse sites, ranging from bog edges to sandy plains to rocky ridges. They also thrive in nutrient-poor ground, growing in acidic, sandy, and rocky soils derived from sandstone, granite, or glacial till. Thus, white pines compete well against hardwood species that are inhibited by poor soil. Favorable conditions for pines abounded in the northern regions of the New England colonies, and the trees grew thickly in the broad river valleys of the Piscataqua and Connecticut rivers, some of them three or four hundred years old. The trunks of these ancient specimens could reach heights of 170 feet and might be as wide as five feet at the base.

Today, Maine is known as the Pine Tree State. Interestingly, however, white pines accounted for only about 5 percent of all trees in the district's vast forests. Nonetheless, explorers and colonists likely saw the species as more abundant than it was. That's because English settlements were concentrated along the large coastal rivers in the southern part of the district, where the pines formed about 10 percent of the region's trees.

The mother country's interest in New England's white pines took time to develop. The problem was distance. To reach the shipyards of England, materials from North America had to travel three times as far as materials from the Baltic ports, forcing a considerable increase in transportation costs. On the plus side, the New England trees were controlled by the king, so harvesting them could end the Royal Navy's dependence on foreign masts. Even better, mainmasts could be made from a single pine, so shipbuilders would not have to construct composite masts. Thus, by the 1650s, England was gradually reducing its need for Baltic trees. While it continued to buy European pines into the 1770s, the forests of New England now accounted for an increasing number of masts on Royal Navy warships.

Even though there were plenty of big pines in the colonies, the task of turning old-growth trees into masts was filled with challenges. First, woodsmen had to fell a pine without shattering the trunk as the massive tree crashed to the ground. To improve the odds, they might cut down a line of smaller trees to create a landing pad, but the tree would still fall with stupendous force. Once the trunk was down, they examined it for defects: the tree had to be sound along its entire length. (By some estimates, nineteen out of twenty trunks were too decayed for use as masts. A flawed tree was still valuable as it could be cut into sections and sent to a sawmill, but it wouldn't have the honor of absorbing French or Spanish cannonballs.) If the trunk passed inspection, the harvesters balked, or hauled, it along a forest track cut to the nearest river, where it could be floated to a coastal port. Because the pines were so tall, the track had to be relatively straight, so the route usually went up and over hills and ridges instead of curling around them. The balking process required between eight and twenty teams of oxen as each trunk could weigh as much as twenty tons. And the job could be dangerous: if braking methods failed, the gravity-propelled trunk could slide down a hill and crush the animals and people in its path.

Clearly, harvesters preferred the shortest possible distances between forest, river, and port. So in the early 1700s, the white pine trade was centered in the Piscataqua River watershed around Portsmouth, New Hampshire, a location filled with broad, pine-rich valleys and an accessible major harbor. As the century progressed, logging reduced the supply of Piscataqua River pines, and the king's men moved to the southeastern coast of Maine and the upper Connecticut River valley. By the 1730s, Falmouth, Maine (today's Portland), had become an important port for pine exports. Regardless of location, the business was lucrative. A single high-quality white pine, 100 to 120 feet long, might be worth £100 at a time when tailor, carpenters, and bricklayers were making £30 to £40 per year.

By the late 1600s, harvested white pines were flowing from New England to the Royal Navy. Unfortunately for the war machine, the colonists were less than supportive. Many chopped down mature trees for their own purposes, such as to clear farmland, rather than preserve them for the king. They also valued it for building. Pine was lightweight but strong: durable, split-resistant, and easy to cut into usable shapes. The trees provided material for the structural framework of houses and bridges as well as for interior finish work such as floors, walls, and cornices. Pine was used for furniture, cabinets, and mirror or picture frames. It could be carved into figureheads for vessels, and it held paint and gilding. Pine boards were an important export commodity New England's trade with the West Indies, often exchanged for molasses.

However, the Royal Navy saw all of this cutting white pines for such frivolous reasons as building houses, fashioning furniture, and opening new farmland as a waste of the king's trees. Clearly, the British government needed to act to reserve the larger pines for masts. The first significant restrictions on harvesting white pines appear in a passage at the end of the 1691 charter for the Massachusetts Bay Colony, which then included much of present-day Maine. It declared that "all Trees of the Diameter of Twenty Four Inches" belonged to King William III and his heirs and successors unless the trees were growing on land "heretofore granted to any private persons"—that is, on land that had been granted or made private before the existence of the charter.[4] While the document did not name a specific tree species, its goal was clearly to reserve large white pines for use as masts. The provision's appearance at the end of the charter may have created the impression that it was an afterthought, but the regulation packed a punch. Cutting one of the king's trees on nonprivate land without a license or approval from the Crown could result in a fine of up to £100 sterling. This amount was close to the value of a mast-worthy white pine on a New England dock, and it far exceeded what a typical laborer or artisan could expect to earn in a year.

The charter's tree-cutting provision marked the beginning of more than eighty years of pine-based conflict between the Crown and the colonists. Over the next several decades, Parliament passed a series of acts that expanded protections for trees used by the Royal Navy. Legislation in 1711 restricted pine felling throughout all of New England as well as in New York and New Jersey. In 1721, "An Act Giving Further Encouragement for the Importation of Naval Stores" added fines for damaging white pines of any size if they were growing on nongranted or nonprivate land. The idea was to save smaller trees for masts far into the future. The act also considered the substantial acreage that did not fall within organized New England townships. To save the pines in these areas for warships, Parliament prohibited the cutting, felling, or destroying of any white pines growing outside the boundaries of an organized township unless the harvester had a royal license to cut trees for masts.

But New Englanders were a clever lot. Even before the 1721 act, colonial governments had been employing a dodge that they would now put to greater use. The strategy involved rapidly establishing new townships in forested areas with few inhabitants. This meant that trees that had been outside the boundaries of a township were now inside them and no longer subject to restrictions on pine harvesting. The creation of these so-called "paper townships" had not gone unnoticed. In 1720, Robert Armstrong, a Crown official in New England, reported to Charles Burniston in England that "the people elude the force of the [1711 act] in reference to masts. They have taken in thousands of acres of the woodland, wherein the best timber grows, and formed the same into their townships," though only a tiny portion of the land is "under any immediate improvement." When the king and council "repealed the [town] grants to keep them from destroying and encroaching [on] his Majesty's timber . . . they continue[d] to make townships and take tracts of land without any title, and the inhabitants cut and fell[ed] all trees at will."[5]

In 1729, Parliament countered these colonial efforts with "An Act for the Better Preservation of His Majesty's Woods." The act included a preamble noting that "great tracts of land, where trees fit for masting grow, have been, in order to evade the provisions of [previous acts], erected into townships." This new act stipulated that "no person or persons . . . shall presume to cut, fell or destroy any white pine trees . . . not withstanding the said trees do grow within the limits of any townships laid out, or to be laid out hereafter in any of the said colonies, without his Majesty's royal license."[6] In other words, the colonists could create all the townships they wanted, but white pines growing "within the limits" of the new township would be subject to the same restrictions as those growing in unorganized land.

By this time, much of the acreage that colonists had viewed as private was now designated as land to be managed under the constraints of the pine laws, which required contracts or licenses before big trees could be cut. But while Parliament could pass acts, laws, and restrictions on white pines for as long as the Thames flowed to the sea, this still left the question of who would carry out its laws? The job of enforcing the pine acts fell primarily to the surveyor general of the king's woods, a position legally established in 1705. That official's first job was to make clear which white pines were to be left for the Royal Navy. Because many felled pines would prove to be unsuitable for masts, the number of reserved trees needed to far exceed the number of ships at sea.

To make these designations, the surveyor general's office dispatched deputy surveyors and marking parties to stamp the king's New England trees with a pattern known as the broad arrow. With a hatchet, the men would cut a three-line arrowhead in a pine's bark, pointing upward, with the two sides of the head at a thirty- to forty-five-degree angle from the center line. The symbol was an old English sign for naval property and was also stamped into items such as Royal Navy cannons. It was a clear and unambiguous message to the colonists: this tree belongs to the king, and you can't touch it without his permission.

For a few colonists, the pine laws were an opportunity to get rich. Men in London bid on Royal Navy contracts to harvest white pines for masts, and those who won the contracts then linked up with agents or representatives in America. These agents received licenses to cut the pines and hired the teams that chopped, hauled, and delivered the massive trunks to the docks. Although the woodsmen earned relatively little money, the agents often made small fortunes from the mast trade. These elites included members of the Wentworth family of New Hampshire and Samuel Waldo of Maine, whose name is now attached to Waldo County, the midcoast region where he owned a considerable amount of land. Because their income depended on tall, intact trees, the agents were delighted to work with the various surveyors general and the Crown to protect the big pines. As I will show, sometimes their connection was very close indeed.

But for the vast majority of New Englanders, the broad arrow was treated as a suggestion rather than a rule. The king might own the trees, but the people owned the saws, and they were ready to use them. The white pine laws prompted years of conflict and defiance as the colonists mostly ignored, subverted, and actively opposed them. After Parliament neutralized the paper townships, some New Englanders took to damaging the trees, setting pine stands on fire or girdling individual trees to kill them. This made them unsuitable for masts but still usable for boards and other purposes. After the colonists transported the white pine logs to a sawmill pond, officials might be lying in wait, ready to confiscate them for being over the two-foot-diameter limit. To avoid this fate, millers would wait till officials were out of sight and then quickly cut the logs into planks, creating a product whose origins were nearly impossible to determine. One sawmill trick was to cut pine boards into widths no greater than twenty-three inches; because a board with a width of twenty-four inches or greater was a dead giveaway that the tree from which it had been cut was at least two feet in diameter. As Robert Albion writes in *Forests and Sea Power*, "a board over two feet wide would be prima facie evidence of illegal cutting, [so] the mills turned them out just within the limit. The inner side of the roofs of old houses of the colonial period will reveal many splendid pine boards, now a rich golden brown with age, twenty-two and twenty-three inches wide, but almost never the damning twenty-four."[7]

Even when the surveyor general did manage to seize and keep illegal logs intact and under his control, millowners would often refuse to buy them at auction, depriving the surveyor's office of an important source of revenue. And from the beginning, these officials had struggled to convict alleged violators in court. Local judges and juries either sympathized with the accused or felt they had more to fear from their neighbors than from the Crown. Again and again authorities made arrests and log seizures, only to watch one defendant after another go free. They complained bitterly to London about these outcomes, and eventually cases involving pine-act violations were moved to local vice-admiralty courts. These courts had originally been established to adjudicate legal matters related to colonial maritime activities, and it was a stretch to use them in white pine cases. But they had the advantage of being juryless so were less vulnerable to local sentiment.

The surveyor general and his deputies had an impossible job. White pines were distributed and poached over a wide region that was sparsely inhabited and difficult to patrol. Occasionally, officials would catch loggers in the act of cutting down a broad arrow tree, but the surveyors were often outnumbered by the woodsmen so had to exercise caution. However, those resisting the pine acts had one major point of vulnerability. While some millers might abet evasion of the pine acts, the sawmills were also choke points in the non-naval use of pine. They were obvious targets for those attempting to enforce the acts, and to catch those thumbing their noses at the king, one could simply watch and raid these operations. But this, too, risked violence, and the few documented attacks on Crown surveyors usually involved conflicts at millponds and sawmills.

For instance, there was the case involving Colonel David Dunbar. In 1728, Dunbar was appointed surveyor general of the king's woods. (He also served as New Hampshire's lieutenant governor from 1730 to 1737.) Dunbar would prove to be about as popular as the diphtheria epidemics that periodically swept through the New England villages. Almost from the start, he antagonized nearly everyone he interacted with, including Jonathan Belcher, governor of both New Hampshire and Massachusetts. Belcher described Dunbar as "the most malicious, perfidious creature that wears a human shape," and this opinion appears to have been widely shared.[8]

Dunbar's grim determination to enforce the pine acts by any means necessary led to a violent confrontation with the colonists of Exeter, New Hampshire. This extensive township contained several millponds and sawmills strung out along the Exeter River, which meandered from its headwaters in the west to its mouth at a tidal river in the village of Exeter. (Some of these mills would later fall within the township of Brentwood when it was incorporated in 1742.) By the early spring of 1734, Dunbar was convinced that this area was a center of illegal white pine harvesting and lumbering, and he intended to make the colonists respect his authority.

But Exeter residents already despised him. On August 21, 1733, Dunbar had spotted a town official, Joseph Thing (or Thyng), on a road near the Pickpocket Mill, and had pressed him to reveal who owned which local sawmills. Thing replied that "he could not tell" because "they were buying and selling almost every day." Frustrated, Dunbar beat Thing with a heavy cane so severely that "he was not well of the blows he received for a considerable time."[9] Then the colonel rode on until he met another local man, John Lusken. When Lusken, too, failed to provide acceptable answers, Dunbar took "violent hold of his hair" and struck him with the same cane. The battering "so broke and bruised his head that a great quantity of blood issued out, and his shoulder was so wounded that he could not work some time."[10] Dunbar would later defend his actions in a letter to Governor Belcher, claiming that the victims "treated me with great insolence which I will never suffer."[11]

After these incidents, Dunbar continued to gather evidence that Exeter sawmills were holding logs from broad arrow trees and cutting them into boards. That may or may not have been true. While it was likely that some locals were illegally harvesting the king's trees, it was also possible that some of the logs Dunbar was monitoring came from private lands that did not belong to the king. Regardless, in late March 1734, he took his case to the vice-admiralty court in Portsmouth, which ruled that the Copyhold and Black Rocks millponds were holding illegal logs. The court decreed that these must be forfeited to the Crown; in addition, about 200,000 board feet of white pine planks sawed from those logs could be seized by Dunbar and his agents in the name of the king.

With the law on his side, Dunbar rode to Exeter in early April and paid a visit to the Copyhold Mill, located about ten miles upriver from the village of Exeter. The mill was owned by a militia major, Nicholas Gilman, and his sons Daniel and Nathaniel. Dunbar informed Major Gilman that he would seize 87,000 board feet of forfeited pine boards after they were separated from non-condemned boards. But how could Dunbar tell condemned from non-condemned? As the citizens of Exeter later pointed out in a petition to Governor Belcher, it was essentially impossible to identify which board came from which log as they "had no sort of Mark nor Indication whereby any man living could discriminate and say such and such parcels were the produce of the Condemned Logs from the others that were the allowed property of private Persons." Any attempt to carry out the court's ruling would be so "Clogged with such Apparent Injustice and Attended with such Chance and guess work, . . . [that] no Wise Judicious Person" would take advantage of the ruling to seize a mill's boards.[12]

However, Dunbar was not known as a "Wise Judicious Person," and he was not moved by such arguments. He would separate the boards, correctly or not; and if he could not sell them to fund his enforcement of the pine laws, they would be burned. It seemed that Gilman had three choices. He could let Dunbar haul away the boards for sale elsewhere, he could buy back what he considered to be his own boards, or he could watch the boards burn. The major chose none of these options. According to Dunbar, he "replied with warmth they should not be burnt on his land [and] the King had no business on his land." The colonel was "a good deal provoked at such disrespectful expressions," but Gilman had home-field advantage. As soon as Dunbar had arrived at the mill, he had been badgered by a crowd of locals, "hallowing, [and] shrieking." Carrying "many small arms from the wood contiguous to ye mill . . . the men . . . kept fireing and hallowing, and running in gangs to and fro as if they would be believed to be Indians."[13] Dunbar assumed that the men were trying to scare him off before he could deal with the forfeited boards, and he was greatly "insulted." Nonetheless, he concluded that the better part of valor was discretion, and he retreated to Portsmouth without taking immediate action.

The colonel wasn't finished with Exeter or the Copyhold Mill, but he knew better than to return to Exeter without support. So on April 22 he impressed a number of Portsmouth men into a surveyor general's task force. Ten of those men would be assigned to separate out the condemned boards and mark them with the broad arrow, and a larger posse would serve as guards as they did their work. Dunbar might have achieved his objective had he not committed the cardinal military sin of dividing his "army" in the face of a numerically superior force. But for some reason, Dunbar decided to split up their arrival time. On April 23, the ten-man marking team sailed up the Squamscott River to the village of Exeter. The surveyor general and his would-be guards followed separately, spending the night in Newmarket, about ten miles away from Exeter, with plans to join the first group the next day.

On the evening of the 23rd, the advance team settled in for the night at a tavern owned by another Gilman, a militia captain named Samuel. At about nine o'clock, as some of the crew sat in the kitchen, three local men "bolted into the room," grabbed a Dunbar man, Robert Galloway, by the hair, and started to pull him toward the door. Fortunately, there was yet another Gilman in the house, a militia colonel named John, who also served as justice of the peace. A Dunbar man, Benjamin Dockum, quickly fetched the colonel, who "commanded the peace," ordered the locals to go home, and "bid" those in the kitchen to go to bed.[14]

Though there had been a "quarrel" (Dockum's word), the crew in the house probably expected that the remainder of the night would be quiet. Most of the men in the kitchen followed Gilman's advice and went upstairs to bed, joining others who had already retired. But peace didn't last for long. At about ten o'clock, between twenty and thirty local men rushed into the tavern. Two of Dunbar's men were still in the kitchen; and when they heard the assault begin, they were able to slip out of the house, blend in with the crowd outside, and avoid detection until morning. Others were not so lucky. Joseph Cross was in another downstairs room when the invaders blew out the candles, chased him through the darkened house, and pulled him outside, where they knocked him down with a club. He begged them not to murder him, but they continued to punch and kick him before finally "bid[ding] him run." He did, hiding behind a fence "till the Riot was past."[15]

Now the rioters charged upstairs to the bedroom where the Dunbar men slept. They burst through the door, extinguished the lights, and shouted, "Now you dogs, we have got you and will be the death of you."[16] Several victims were dragged down the stairs and out of the house, where they were clubbed and beaten with fists. One man later testified that he "was in great danger of his life, having received several wounds, and lost a great deal or blood."[17] After the beatings, most of the men were allowed to limp away as best they could. However, one unfortunate, Joseph Miller, was dragged to the riverbank "where was a pile of boards over which they threw him and down the bank about fifteen foot by which he received a great hurt in his back, where he lay till the next morning being afraid to be seen again least he should be murthered [murdered]."[18]

There were acts of kindness and courage in midst of the chaos. After Benjamin Dockum was dragged out of the tavern and beaten, a man "step't in and took hold of [me] and said run if you can and I will help you, at which I, with his assistance, got off a little way, and hid myself under a wharf." Later, when the tide rose, Dockum "crawled out and lay under a pile of boards 'till daylight."[19] He later found that he had lost his "pocketbook," which had contained twenty or thirty shillings. Benjamin Pitman was also dragged from the upstairs bedroom but, once outside, managed to break free and run to the house of Henry Marshall, a hatter. The Marshalls were in bed when the terrified man burst in, but they quickly decided to help, hiding him under a bedspread. Marshall later testified that Pitman was soon followed in by a mob. The hatter "denyed that he was there, they then threatened him, but their voices seemed disguised [or perhaps Marshall was afraid to identify them], and they continued around the house to our great terror."[20] Apparently the rioters didn't find Pitman, who remained safe for the rest of the night.

As dawn broke on the morning of April 24, Dunbar's men emerged from hiding. They were dazed and sore, but all were still alive. It was time to retreat, but the river, the obvious route, was impossible: their boat had been cut to pieces, the sails torn or stolen. They would have to hobble down the road toward Portsmouth. As they walked, they met Colonel Dunbar and the posse of guards. On learning what had happened, Dunbar became enraged. Riding into Exeter, he demanded to see the three justices of the peace: John and Nicholas Gilmore and Bartholomew Thing. Dunbar already had a low opinion of Thing; in a letter to Governor Belcher he had compared the militia major to "the late [Jonathan] Wild of London, who neither robbed or stole himself, but was in confederacy with a 1000 thieves and robbers."[21]

All of the justices said they'd known nothing about the night's events until nine o'clock the next morning. Dunbar did not believe them, describing their responses as "sanctified fizzes," "scarce credible when the outcry of murder was heard more than a mile from the town."[22] It is possible that they were very sound sleepers, even John Gilman, who had been in the tavern's kitchen an hour before the second outbreak of violence. Or they may have been lying. Whatever the truth, Dunbar knew they were unlikely to change their stories. However, he did order them to take statements from the wounded men, mostly likely hoping to use them in future prosecutions. There was little else he could do. Half of his force was battered, his boat was out of action, and he was surrounded by Exeter men who had taken part in the fray. It appears that he returned to Portsmouth without his boards.

But Dunbar's anger had not abated. On April 29, he was firing off angry letters to Governor Belcher about what was becoming known as the Exeter Riot. Belcher hated Dunbar, but he could not completely ignore an attack on men in His Majesty's service. In early May, he issued a proclamation describing the crimes committed in Exeter and calling on local judges, justices of the peace, sheriffs, and constables to seek out the names of the men involved. It also offered pardons to those who would reveal the names of their accomplices. Much to Dunbar's disgust, however, it did not offer cash rewards or additional incentives: clearly, Belcher's proclamation was just for show. Moreover, later that month, he accepted a petition from Exeter residents in which they complained about Dunbar's actions, including the August canings and his threats to burn what they saw as their own boards. Belcher happily passed this petition on to officials in London, adding his own list of grievances against Dunbar for good measure. In the end, there were no arrests or convictions related to the Exeter Riot.

In 1743, New Hampshire's governor, Benning Wentworth, assumed the title of surveyor general of the king's woods, thus solving the problem of disagreements between surveyor and governor. Wentworth was motivated by more than just a desire to serve the growing British Empire and King George II: his family members were making small fortunes as agents in the white pine mast trade, especially his brother, Mark Hunking Wentworth. Now, as surveyor general, Benning would be in an excellent position to advance his family's interests because the surveyor general had jurisdiction over all of the New England colonies, not just New Hampshire.

The Wentworths had made much of their fortune by cutting pine for masts in the forests of the Piscataqua watershed. However, as I've mentioned, by the middle of the eighteenth century, the pine stock in this region had declined severely, and the family recognized that the trade's future lay in the valley surrounding the upper Connecticut River. Today, this section of the river forms the border between New Hampshire and Vermont, but at the time all of the river land was claimed by the New Hampshire colony, although the New York colony disputed that claim. So, in the mid-1700s, the Wentworths set out to control vast tracts of upper valley forest, both through direct purchases and by employing a neat trick: Governor Wentworth reserved a portion of land for himself in each of the new townships he granted along the Connecticut. Thus, he had a vested interest in administering the white pine laws along the river, which is why, in his capacity as surveyor general, he focused his enforcement efforts in the region.

Like Dunbar, Wentworth began by focusing on millponds and sawmills as hubs for marking and seizing massive logs cut from broad arrow trees.[23] But he had a lot of area to patrol: the Connecticut is New England's longest river, stretching from the Canadian border to Long Island Sound. So, in 1753, he hired Daniel Blake as deputy surveyor and sent him down to the Connecticut Colony with a commission to protect the white pines of His Majesty's Woods—which also happened to be full of Wentworth family pines. Blake was to seize illegally cut trees and their logs and mark them with the broad arrow. As an incentive, he would receive an allowance of all forfeits and seizures.

However, Wentworth's plan soon went awry. In the spring of that year, Blake rode to a millpond near the riverside village of Middletown, Connecticut, perhaps hunting for trees and logs that had floated downstream from the northern colonies. There he was spotted by Daniel Whitmore, scion of a family of lumbermen and sawmill owners. In Whitmore's eyes, Blake was a very unwelcome guest; and before the deputy surveyor could assess the legality of the mill's activities, Whitmore threw him into the pond. In a letter to Governor Roger Wolcott of Connecticut, Wentworth complained about the dunking, claiming that Blake's "life was much endangered, & he [was] otherwise disabled there by from pursuing the Kings business." Wentworth expected Wolcott to deal with the matter by "carrying on a prosecution against the said Daniel Whitmore."[24]

But Wentworth and Blake were disappointed: Daniel Whitmore was never brought to trial. As an elected official, Governor Wolcott had nothing to gain from prosecuting a Connecticut citizen in a case involving a deputy sent by a royal official in New Hampshire to enforce the very unpopular pine acts. For a time, Blake persisted. In 1759, he petitioned the Commissioners for Trade and Plantations in London for compensation for "several hardships and personal injuries sustained by him in the execution" of his office. However, he was again disappointed. The commissioners concluded "that the injuries and hardships complained of were of a private nature, and that it was not within their power . . . to give any relief."[25]

In 1767, in part due his dubious land dealings in the disputed territories west of the Connecticut River, Benning Wentworth was replaced as both governor and surveyor general. Both positions were awarded instead to his nephew, John Wentworth, who took his job as surveyor seriously. He traveled extensively in the forests, developing a deep knowledge of the wilderness, talking to the woodsmen he met, and earning their respect. While Wentworth did enforce the pine laws, he also tried to negotiate directly with those who were affected by them. He soon concluded that it was impossible to mark and protect every big white pine in the American colonies. Instead, he proposed that small reservations of large pines be set aside to supply the Royal Navy with masts. Cutting for lumber would not be allowed within the reservations, and there would be severe penalties for violators. Outside of these patches, however, the colonists would be free to do as they pleased. Unfortunately, because Wentworth's tenure was ended by the American Revolution, most of his ideas weren't put into action.

While surveyor general's efforts did reduce tensions between the local timber industry and the tree police, he was still a royal official enforcing unpopular laws; and in the charged atmosphere of the early 1770s, it didn't take much to trigger a clash between Crown and colonists. In the winter of 1771–72, a deputy surveyor named John Sherburn (or Sherburne), traveled to the Piscataquog River valley in recently organized Hillsborough County, New Hampshire. The region's industry centered around white pines, and most of that material was transported via tributaries of the Merrimack River, including the Piscataquog. Sherburn discovered harvested pines logs with diameters greater than twenty-four inches at several millponds and sawmills in Bedford and Goffstown as well as 270 suspect logs at Clement's Mill in the village of Weare. Some of the logs at Clement's were as large as thirty-six inches in diameter. The vice-admiralty court at Portsmouth quickly ruled that the logs had been produced by illegal cutting, and a notice in the February 21, 1772, edition of the *New Hampshire Gazette* announced that anyone "claiming Property in the following WHITE PINE LOGS seized by Order of the SURVEYOR GENERAL in Goffstowne and Wear, . . . may appear at a Court of the Vice Admiralty to be held at Portsmouth on Thursday, [February 27], . . . and shew Cause why the same should not be decreed Forfeited."[26]

Given the size of the logs, it would be hard to argue that they were not from broad arrow pines. Nonetheless, the millowners in Bedford and Goffstown hired Samuel Blodgett, a lawyer from Goffstown, to argue their case in court. This decision backfired. According to the historian William Little, "the governor fell in love with [Blodgett] at first sight, won him over to his side, and . . . made him a deputy 'Surveyor of the King's Woods.'" In his new role, the lawyer was now commissioned to "preserve the King's woods from trespass or waste" and required to "prosecute and punish" violators.[27] This was probably not what Blodgett's clients had in mind when they retained him. Nevertheless, they eventually agreed to a deal: they would pay a fine but retain their logs for milling and sale, and their cases would be dropped.

The men of Bedford and Goffstown had settled their cases. But the men of Weare had not. Among them, the millowner Ebenezer Mudgett (or Mugget) was seen as the chief offender, and the court issued a warrant for his arrest. Perhaps if Mudgett were detained, the other Weare men would fall into line and pay their fines. So on April 13, 1772, Sheriff Benjamin Whiting and Deputy John Quigley rode to Weare, located about fifteen miles southwest of Concord, to serve a warrant on Mudgett. After they informed the millowner that he could pay a fine for the forfeited logs or be arrested, he asked if he could meet them in the morning with bail money so as to avoid a trip to the jail in Exeter. The sheriff and his deputy had no objections, and all agreed to convene in the morning at the Quimby Inn.

As Whiting and Quigley settled into the Quimby for the night, Mudgett gathered his friends and relations. Then, at dawn, he walked into Whiting's room and announced that his bail was ready. According to an April 24, 1772 article in the *New Hampshire Gazette*, Whiting replied that "he need not have hurried quite so much, but [could] have stopped till it was lighter. . . . However, . . . he would get up, and wait on him."[28] But as the sheriff was dressing, ten or twenty local men with blackened faces rushed past Mudgett into the room. Whiting was carrying pistols but was unable to get off a shot before the invaders pinned, clubbed, and whipped him. Deputy Quigley showed more fight but ultimately suffered the same fate. When the Weare men were satisfied that the beatings had gone on long enough, the victims were ordered onto their horses, with tails and manes now sheared, and driven out of town.

But Whiting was tough. He was determined to return to Weare, this time with more than a single deputy. He consulted with two local militia colonels, John Goffe of Manchester and Edward Lutwytche of Merrimack, who gathered their regiments and marched to Weare. The posse found some of the ruffians in the woods; and though most of the men were able to flee, one was caught and apparently named names. Mudgett and seven others were indicted as rioters, routers, and disturbers of the peace. They were charged with assault, in which they "beat, wounded and evilly intreated so that his life was despaired of, and . . . [did] great damage of the said Benjamin Whiting, and against the peace of our Lord the King his crown and dignity."[29]

The evidence was overwhelming. The men pleaded guilty and were each sentenced to pay fines of £1, the equivalent of roughly $500, a relatively light punishment, given that they'd beaten a sheriff. To some degree, the men of Weare had gotten away with directly and violently challenging Crown officials. The event became known as the Pine Tree Riot, and the strategy of blackening faces while resisting the king's authority would be used again in December 1773, during the Boston Tea Party. As Weare chronicler William Little argues, "The only reason why the 'Rebellion' at Portsmouth and the 'Boston tea party' are better known than our Pine Tree Riot is because they have had better historians."[30]

The white pine acts may have had a limited impact on the preservation of old-growth pines for the Royal Navy, but they did a very good job of teaching New Englanders how to resist unpopular royal decrees—skills that played a key role in the war for American independence. Moreover, because the acts affected citizens across classes and occupations—including lumbermen, sawmill operators, merchants, colonial legislators, and judges— acts of resistance created a unity that would become valuable during the events of the 1770s. It's no surprise that the white pine was displayed on several flags during the Revolutionary War.

Chapter 3

Boundary Disputes and the Aroostook War

O N FEBRUARY 19, 1839, Governor John Fairfield sent a letter
from Augusta, Maine, with instructions that it be delivered
to Major-General Sir John Harvey, the British lieutenant-governor
in Fredericton, New Brunswick. Fairfield was responding to an
earlier letter from Harvey, who had written to inform the governor that a party of armed Maine men, then positioned along
the Aroostook River, would have to leave. Harvey's position was
clear: if the Mainers didn't leave, he would send in a force of
British regulars and New Brunswick militia to push them out. In
his reply, however, Governor Fairfield made it clear that Maine
men would not budge:

> I have no threats to make, no boastings to indulge. If Maine
> does her duty, as I trust in God she will, nothing that I
> could say in advance would add to the glory of her career.
> If she proves recreant to her duty, and tamely submits to
> be expelled from her territory by a force that she could
> successfully resist, nothing that I can say would tend to
> diminish the measure of her ignominy and disgrace.[1]

American and British forces were about to clash, and a third
war between the two counties would likely follow. But what had
led to them to this dangerous moment? The issue was a dispute
over a poorly defined border—and white pines.

Pines' economic importance had not ended with the American

Revolution, and the question of who would profit from them remained. In northern Maine and southeastern Canada, the answer hinged on the murky position of the northern border between the United States and the British Empire. Even as late as 1839, no one knew for certain where that border was located. The problem stretched back to the 1783 Treaty of Paris. At that time, the northern Maine wilderness was poorly mapped, and diplomats had neither the survey data nor the geographic knowledge to define a clear border. Inevitably, they produced descriptions that were vague and open to interpretation. By the terms of the treaty, surveyors were supposed to find the source of the Saint Croix River and map out a line due north to "highlands which divide those rivers that empty themselves into the River St. Lawrence from those which fall into the Atlantic Ocean." The boundary was meant to follow "said highlands" to the west and southwest "to the North Westernmost head of the Connecticut River."[2] But there were major disagreements about the position of the various geographic features described in the treaty, and different answers to these questions led to different conclusions about the international border's location.

In the decade after the Revolutionary War, issues linked to the unsettled border created tensions between Britain and the United States. In the 1783 treaty, the boundary lines had been based, in part, on the Saint Croix River, so now the two countries squabbled about the exact location of the river's source. They even disagreed about which river was the true Saint Croix, given that multiple rivers known by multiple names empty into Passamaquoddy Bay, which abuts both Maine and Canada. In 1798, a bilateral commission resolved the issue, in part, by turning to a 1772 survey of de Monts's and Champlain's ill-fated, scurvy-ridden Saint Croix Island colony, which had historically marked the mouth of the river (see chapter 2).[3] Now, with the eastern boundary of Maine established, the next and more difficult step was to draw the boundary between the district (then still part of Massachusetts) and the British lands to the north and northwest.

This was a challenging task, and little action was taken until the 1814 Treaty of Ghent, which not only ended the War of 1812 but also tackled border ambiguities stretching from the Atlantic to the Great Lakes.

According to this treaty, commissioners would initiate surveys to gather geographic data along the disputed line and map the border, thereby "finally and conclusively [fixing] the said boundary."[4] However, there was a catch: the boundary would be fixed only if both commissioners could agree about its final location. To that end, all parties would need more data. So in 1817, the British surveyor Colonel Joseph Bouchette and the American surveyor John Johnson began what would become a multiyear project, starting their work at the monument marking the source of the Saint Croix River. At this time, the monument was a yellow birch tree growing in swampy ground, made distinguishable from other trees by the iron hoops that surrounded it and by a stake marked "S.T. 1797," apparently signifying the surveyor Samuel Titcomb and the year the stake had been planted. Bouchette and Johnson thought they needed something better than a birch tree to mark such an important spot, so they and their crews built a new monument out of a cedar log, cut to twelve feet in length and hewn to eight inches square. They carved "New Brunswick July 31, 1817" and "United States July 31, 1817" on the eastern and western sides, respectively, and carved Bouchette's and Johnson's names on the northern and southern sides, respectively.[5]

Egos satisfied, the surveyors headed due north to find "those Rivers that empty themselves into the River St. Lawrence"— that is, the boundary of the Saint Lawrence River's watershed. Advancing winter chased them out of the wilderness before they could reach the watershed; but they tried again the following year, and on September 2, 1818, they reached a stream they named the Beaver River, which flowed northwest in the direction of the Saint Lawrence. They had found and crossed the elusive "highlands which divide those rivers that empty themselves into the river St. Lawrence from those which fall into the Atlantic Ocean." At

FIGURE 2. Monument at the source of the Saint Croix River, erected 1817. Note which side of the monument faces the viewer. The lithograph is from Joseph Bouchette, *The British Dominions in North America* (London: Colburn and Bentley, 1831).

this point, the survey teams were between sixty-five and seventy miles north of the Saint John River and about 145 miles from the source of the Saint Croix. Now all they had to do was follow the highlands marking the rim of the Saint Lawrence watershed to the headwaters of the Connecticut River; this would determine the international border. They completed that task in subsequent years, and it appeared that a border could now be drawn.

This boundary line would have placed the entire Saint John River watershed west of the line running due north from the Saint Croix under U.S. control, including all land draining the Allagash, Fish, Aroostook, and Madawaska tributaries. It also would have made Maine about 5,000 square miles larger than it is today. The Americans liked this line, but the British did not. If they had accepted the Saint Lawrence highlands line, they would have had much less white pine to harvest and much less land to clear and farm. In addition, for a stretch of more than one hundred miles, the British would have possessed only a narrow strip of land between the Saint Lawrence River and the United States. This would have allowed the Americans to station troops on highlands that were uncomfortably close to Quebec City; and given that the Americans had just invaded Canada during the War of 1812, this was not a minor concern. Finally, a section of a critically important military road, known as the Halifax Road or the Grand Communications Route, would have now passed through American territory. This thoroughfare connected Quebec City on the Saint Lawrence River to Saint John, New Brunswick, on the Bay of Fundy and ultimately to Halifax, Nova Scotia, on the Atlantic Ocean. The road was essential for provisioning Quebec (then known as Lower Canada) and Ontario (Upper Canada) when the Saint Lawrence was frozen, and its value had become clear in the recent war.

The British commissioner was certain that his nation's diplomats at the Treaty of Paris peace talks would not have approved of such a border. And, in fact, the British saw a way out. The 1783 treaty had not specified the distance to "said highlands" or even clearly defined the meaning of "highlands." Thus, there might be multiple ways of mapping the treaty's provisions onto the topography of the region. The British argued that the highlands around the Beaver River were too far away to be the highlands described in the treaty. In addition, they just weren't high enough in elevation to count as highlands. The American-favored line had to be wrong.

To support their claims, British surveyors needed to find another location for the "highlands," someplace much farther to the south, and the crew knew exactly where to look. In late September 1819, starting at the Saint Croix's source and following a line due north for about forty miles, surveyor William Odell arrived at Mars Hill, an isolated, 1,600-foot-high mountain just west of the line, the first "highland" on the way to the Saint Lawrence and about a hundred miles closer to the Saint

Croix than the Beaver River site. From the top of Mars Hill, Odell looked west and spotted "a range of mountains very high, and apparently bald."[6] Odell's "range," which included massive Mount Katahdin, consisted of a broken chain of mountains and ridges running west for more than a hundred miles to a tributary of the Chaudière River that drains into the Saint Lawrence. To return to the language of the 1783 treaty: these highlands ran to a western terminus where "those rivers that empty themselves into the river St. Lawrence" were divided from "those which fall into the Atlantic Ocean." They also separated the Penobscot River watershed to the south from the Saint John River watershed to the north. If the boundary were to follow Odell's range, the British could claim the entire length of the Saint John tributaries, among them the Allagash, Fish, and Aroostook rivers, along with Telos and Chamberlain lakes. That would add up to a lot of white pines.

Of course, the Americans were not pleased. This boundary line would have cut off the top of present-day Maine, subtracting about 7,000 square miles from its area. So they were quick to point out problems with the British-favored border. The Treaty of Paris required that the eastern end of the northern border be located at a point in the highlands where the rivers "empty themselves into the River St. Lawrence." But the eastern, or Mars Hill, end of the British line was about a hundred miles south of any ridges where rivers flow northwest into the Saint Lawrence. In addition, the British boundary would have assigned all land in the Saint John watershed to the empire, even though none of the runoff in this watershed flows into the Saint Lawrence, as the treaty required. In fact, the Saint John empties into the Bay of Fundy, which leads to the Atlantic Ocean.

But the British had an answer. Conveniently, they viewed the Bay of Fundy as separate from both the Saint Lawrence River and the Atlantic Ocean, and they noted that it was not explicitly referenced in the Treaty of Paris. By ignoring the rather obvious connection between the Atlantic and the Bay of Fundy, they could claim that the treaty did not unambiguously award the United States the watersheds of rivers flowing into the bay. Therefore, the Saint John watershed might be included in the British claims.

By the early 1820s, the surveying work was finished, and the commissioners needed to map the final boundaries. Clearly, however, the American and British commissioners would never agree on the location of Maine's northern and western borders. Their conflicting claims were literally a hundred miles apart, and the 8 million acres between the two proposed boundaries had become known as the Disputed Territory. If this land had remained underpopulated, hard to reach, and low in value, then the treaty's vagueness and the commissioners' disagreements might not have been such a burning problem. But as timber-men decimated pine stands in one watershed after another, the center of white pine logging in Maine had shifted. In the 1700s, pine harvesting had been centered in the south, along the Saco and Presumpscot rivers. By the early 1800s, it had shifted east to the Androscoggin and Kennebec rivers and, as the century progressed, had moved even further east to the Penobscot River and north to tributaries of the Saint John. Along those tributaries, almost all of the pines worth big money were growing in the Disputed Territory. Now it really mattered who had the right to sell the land or charge loggers for timber-cutting rights.

Until the boundary dispute could be resolved, neither Britain nor the United States could claim uncontested jurisdiction in the Disputed Territory. But at various times in the 1820s and 1830s, the province of New Brunswick, the states of Massachusetts and Maine, and both national governments would act as if they controlled all or part of this legal no man's land. Conflicts over jurisdiction began before the boundary surveys were finished. As early as 1818, Nathan Baker, who was from the Kennebec River area, traveled north and settled with several other families on land along the Saint John River near the British settlement of Madawaska. No doubt the old-growth white pines growing along river had played a role in bringing Baker to this remote location

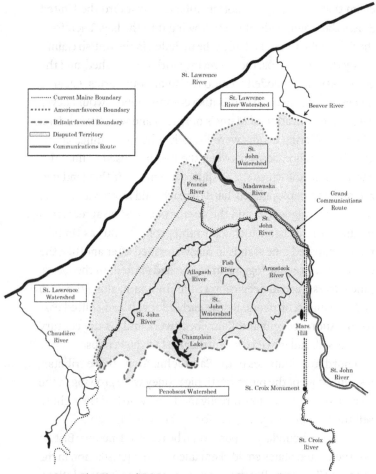

FIGURE 3. The Disputed Territory.

because, by February 1819, he had "ten or twelve hundred tons of timber [about 300,000 board feet] . . . upon the banks of the river St. John, on the north side," ready to be floated for sale down to the port of Saint John, New Brunswick.[7]

Provincial officials were not pleased. When word of this intrusion reached Charles Bagot, the British ambassador in Washington, D.C., he consulted with his counterpart, Secretary of

State John Quincy Adams. In a letter to Colonel Thomas Barclay, the British representative on the boundary commission, Bagot stated that "Mr. Adams appears to think that the persons referred to [Baker and his fellow settlers] . . . are, in reality, what are called squatters, and must be dealt with accordingly."[8] While this may not have been Adams's intent, the British took his response to Baker's action as unofficial acknowledgment that the British government had more authority in the Disputed Territory than the Americans did, at least until the international boundary issue was straightened out.[9] This assumption would lead to trouble.

In 1820, Maine separated from Massachusetts to form its own state, and the value of the Disputed Territory's white pines increased in significance. To fund its operations, the new state government desperately needed the revenue generated by selling both wilderness timberlands and logging rights on public acreage to settlers and investors. To turn pines into public cash, the state used a grid system to divide a large portion of the Disputed Territory south and west of the Saint John River into rows and columns of unorganized townships, usually in the form of six-mile-by-six-mile squares, each containing about 23,000 acres. Massachusetts was awarded half of these unorganized townships as compensation for the loss of the district, which was why the Commonwealth continued to have a stake in land sales and timber rights in the Disputed Territory. New Brunswick did not want to be left out of the fun; and in the absence of a settled boundary, the province felt free to issue its own land grants and timber-cutting permits for pines growing in the Aroostook and upper Saint John valleys.

In short, all three jurisdictions were trying to establish the right and authority to make money from tracts whose ownership remained highly debatable. As a result, when Americans harvested pine on contested land, New Brunswickers called them trespassers, and vice versa. Even Maine and Massachusetts didn't fully trust each other, and they sent dueling land agents into the woods to keep an eye on the competition's timber practices. At one point, Maine officials worried that loggers with Massachusetts permits might cut down white pines in a Maine township and then claim that they'd been cut in an adjacent Massachusetts-controlled township, thus bypassing the payments due to Maine.[10]

To address these problems, Maine, Massachusetts, and New Brunswick pledged in 1826 to stop selling timber-cutting permits in the Disputed Territory until the border issue was resolved. However, words were not enough: it soon became necessary to add an enforcement component. So in the late 1820s, New Brunswick, still assuming that Britain was the chief authority in the region, appointed magistrate James MacLauchlan as warden of the Disputed Territory, giving him the power to stop illegal timber cutting and seize all illegal timber that he could find. According to MacLauchlan, his biggest problem was American loggers who believed that their right to harvest trees had been granted by Maine and Massachusetts land agents. But the warden was determined to stop all illegal activity, whether the woodsmen were Americans or Canadians. As a result, the Americans were generally willing to accept his oversight and to turn to him with their own allegations of trespassing and prohibited logging. Although the illegal harvesting and permit feuds never completely ended, MacLauchlan did have some success at curbing these practices in the Disputed Territory. His job became easier when the Panic of 1837 triggered an economic depression and reduced the demand for pines. But as the depression eased, their value rose again, and so did the potential for conflict.

In the late 1830s, merchants in New Brunswick began sending more and more settlers and loggers into the Disputed Territory to harvest the increasingly lucrative white pines. Government officials in Maine took notice, and they quickly became concerned that such trespass timbering would deprive the state of essential public revenue, which it needed to reduce the massive debts it had accrued during the panic. Toward the end of 1838, land agents for Maine and Massachusetts authorized George Buckmore to keep an eye on the New Brunswickers who were logging pines in the Disputed Territory; and in December 1838, he submitted a report to a Maine land agent, Elijah Hamlin, with explosive news. Buckmore estimated that, by winter's end, the amount of trespass timber would be ten to twenty times greater than the previous year's. Later, George Coffin, a land agent for Massachusetts, would conclude that Buckmore's estimates had been greatly exaggerated. But it was too late: fear and anger about British logging had already become fixed in Maine minds.

Hamlin forwarded Buckmore's report to Governor Fairfield, along with his own warning about extensive damage to public lands He urged the state to take urgent action to stop the New Brunswick loggers, which in his view would require at least fifty armed men. Fairfield had to act. In January 1839, he shared the report with the state legislature, emphasizing that, if nothing were done to stop the loss of trespass timber, Maine would lose a significant amount of money. The governor called on legislators to authorize the formation of a posse that would be placed under the command of the state's land agent. The posse would proceed to the Aroostook and Fish valleys, where it would "seize the teams and provisions [of New Brunswick loggers], break up the camps, and disperse those who are engaged in this work of devastation and pillage."[11] The legislature quickly passed a resolve that put Fairfield's recommendations into action. The Aroostook War had begun.

Fairfield assigned a new land agent, Rufus McIntire, to lead the posse that charged with expelling the New Brunswickers from their camps in the Aroostook Valley—an area that Mainers strongly believed was Maine land. McIntire was a veteran of the War of 1812, a former U.S. congressman, and no fan of the British Empire. The task of organizing the posse fell to Hastings Strickland, the sheriff of Penobscot County, which at this time extended north beyond its present-day boundaries to what the state believe was its true border. In early February, the posse's two hundred men left the Bangor area for a base camp at Masardis on the upper Aroostook River. Although temperatures

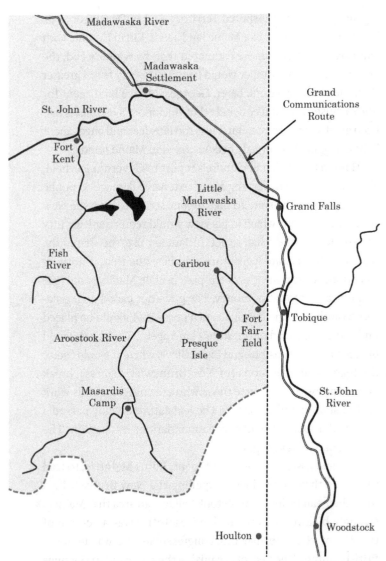

FIGURE 4. Sites of the Aroostook War. Some town names mark modern towns not present in 1839.

were well below zero, winter conditions meant that men and supplies could travel by sleigh over a track that might be impassable during thaw.

Downstream, the New Brunswick lumbermen knew that the Maine posse was heading their way; and when the Maine men arrived at the timbering site, they discovered that the operations had been abandoned. The loggers were probably unwilling to risk their lives for their faraway merchant employers. But the posse kept up the pursuit; and on February 11, near the present-day town of Presque Isle, it caught up with fifteen or twenty armed men trying to hurry their horse teams downriver into New Brunswick. The armed men formed a line, and one snapped off a shot at Sheriff Strickland as he charged them. The bullet missed, only slightly wounding his horse, and this appears to have been the extent of the gunfire.

Outnumbered by the posse, the New Brunswick loggers and their teams fled down the river, but they were soon captured and subjected to an impromptu legal proceeding. The "court" charged them with trespassing on Maine public lands and shooting at Strickland. Five were detained and sent south to the Penobscot County Jail in Bangor. The posse then continued, without opposition, downriver to the mouth of the Little Madawaska, near today's town of Caribou. So far everything had gone swimmingly, but Agent McIntire had more in mind. He wanted to talk to Warden MacLauchlan. His plan was to offer to release the illegal timber to the warden if the trespassing loggers would agree to pay a fee of five shillings per ton. While the posse had been sent to assert Maine's unofficial authority in the Aroostook Valley, the agent understood that it was just as critical to turn already cut pines into state revenue.

McIntire knew that MacLauchlan was currently sixty miles away at the Madawaska settlement, but he thought they could arrange to meet in a day or two at the home of a mill operator named James Fitzherbert. Perhaps lulled by success, he decided to separate from the posse and travel with two other Americans to Fitzherbert's house, about six miles downstream from the Little Madawaska, leaving the remaining men behind to mark the seized timber. McIntire was now only two or three miles from the New Brunswick border.

What he didn't know was that, across the border, fifty or sixty New Brunswick lumbermen had gathered along the Saint John River, a few miles downstream from the mouth of the Aroostook, at the Tobique settlement home of Benjamin Tibbets (near present-day Perth-Andover, New Brunswick). When they learned that McIntire was isolated at Fitzherbert's, they took action. Led by Asa Dow, a local farmer and lumberman whose crews were cutting white pines along the Aroostook, a company of fifteen or twenty armed men traveled by sleigh to Fitzherbert's, where they surrounded the house and arrested McIntire and a few other Americans without bloodshed. By February 14, he had been transported to Fredericton, where he was charged with interfering with the Disputed Territory and acting as if he had jurisdiction along the Aroostook River.

After McIntire's capture, the Maine posse had two options. They could charge across the border into New Brunswick to rescue their leader. Or they could scramble back up the Aroostook River to their base camp at Masardis. They chose to scramble. Apparently, the New Brunswick loggers had spread rumors that a force of up to "300 White men and 25 Indians, well-armed" was on their way to drive the posse out of the lower valley.[12] The alarmed posse abandoned their camp at the Little Madawaska so rapidly that they left some of their sleighs behind.

As the company began to fortify themselves at Masardis, Sheriff Strickland tore back to Bangor, covering 160 miles in less than two days. Arriving on February 14, he then flew on to Augusta to tell the governor about McIntire's capture and the danger of an attack at Masardis. Some observers thought that the sheriff's speed may have been inspired by more than just a desire to transmit information. A few writers for the Bangor and Portland newspapers suggested that he was trying to get away from that rumored New Brunswick attack force as fast as possible, but their comments may have been politically motivated, given that the position of sheriff was an elected one.

Meanwhile, on February 13, Warden MacLauchlan arrived at Tibbets's home in the Tobique settlement and learned about the events of the past few days. He was not sure that McIntire's capture had been a good idea, but the prisoners were already on the road to Fredericton. So the warden turned his attention to the posse that had retreated into the upper Aroostook River valley. Accompanied by two other men, and perhaps supported by his confidence in own his authority, he proceeded upriver to the fortified camp at Masardis. In the absence of McIntire and Strickland, Stover Rines, the captain of the posse's Bangor contingent, was in command at the camp. MacLauchlan told Rines that no New Brunswick force was preparing to attack the Maine men and assured the captain that he opposed McIntire's arrest and detention. Still, he made it clear that the posse had no authority or jurisdiction in the Aroostook River valley, and he told the company to disperse. Rines and his compatriots disagreed. Instead, they arrested the warden and his companions, in part as retaliation for McIntire's arrest, and on February 15 packed them off to Bangor. According to the February 26, 1839, edition of the *Portland Advertiser*, MacLauchlan was "brought to the city a prisoner, and is now at the Bangor House."[13] This new hotel, completed in 1834 at the behest of wealthy lumbermen, was intended to convey the message that the city had class and sophistication. It's not clear if MacLauchlan was awed by the hotel's attempts at elegance; but, as jails go, it could have been worse.

At this point in the Aroostook War, the score was tied. Each side was holding one major player and a few minor players from the opposing team, and so far no one had been injured or killed. But things could get worse, and the possibility of another full-scale war between America and Great Britain could not be ignored. The Grand Communications Route was so valuable to the British that their military would certainly fight to retain at least that portion of the Disputed Territory north of the Saint John River. While the American president, Martin Van Buren, had no interest in a war with Britain, he also could not easily surrender 12,000 square miles of disputed land to the nation that had burned down the White House just twenty-five years earlier. In addition, it was an open question as to whether the federal government could give up part of a sovereign state's land without that state's permission, and Maine was determined to retain at least the land south of the Saint John because of its crucial pine-related revenue. At the local level, residents of both New Brunswick and Maine knew that access to old-growth white pines would improve their personal finances. And so, in the second half of February 1839, words and troop mobilizations escalated.

On February 13, shortly before McIntire and his captors reached Fredericton, Major-General Sir John Harvey, the lieutenant-governor and commander-in-chief of New Brunswick, issued a proclamation stating that an armed posse from Maine had "invaded a portion of this Province under the jurisdiction of Her Majesty's [Queen Victoria's] Government." Harvey made it clear that he considered the Aroostook River valley to be under British administration and that the posse's actions constituted an illegal raid on the queen's territory. However, he also criticized the response of the New Brunswick loggers led by Asa Dow, stating that they had acted "without any legal authority" and had "broken open certain Stores in Woodstock [downstream from the Tobique settlement], in which arms and ammunition belonging to Her Majesty were deposited." Yes, the company under Dow was responding to an American incursion, but there was a proper British way to do things and Harvey would not allow unauthorized bands of New Brunswickers to operate in his province. Instead, it was the responsibility of the provincial government to "adopt all necessary measures for resisting any hostile invasion or outrage that may be attempted upon any part of Her Majesty's territories or subjects." He ordered British military forces into the region to repel any invasion from Maine but also to "prevent the illegal assumption of Arms by Her Majesty's subjects."[14]

On the same day, Harvey sent a letter to Governor Fairfield, emphasizing that Britain had jurisdiction in the Aroostook River valley because "it has been agreed betwixt the two General Governments" that this land "shall remain in the exclusive possession and jurisdiction of England until that claim can be determined." He expressed "the utmost surprise and regret that . . . an armed force from the State of Maine has entered the [disputed] territory" and declared that his instructions from the Crown did not permit him "to suffer any interference with that possession and jurisdiction" until the boundary question had been settled. Harvey regretted that Maine had put him into the position of choosing between "failing in his duty, by abstaining from the fulfilment of the commands of his sovereign, or, by acting up to them, placing the two countries in a state of border collision, if not the two nations in immediate and active hostility." Why had Fairfield felt the need to make Harvey's life difficult? Did the Maine governor really want to start a war over pine trees?[15]

Harvey made it clear that the Maine posse would have to withdraw from the Disputed Territory; if necessary, he was ready to use military force to occupy the Aroostook region and compel the men to leave. But he ended his letter on a more conciliatory note. He assured Fairfield that he understood Maine's concerns about "timber spoilation," and he offered a carrot. He told the governor that he had "given directions for a boom to be placed across the mouth of the Aroostook [which was indisputably within New Brunswick], where the seizing officer, protected by a sufficient guard, will be able to prevent the passage of any timber into the St. John in the spring." In other words, the Maine posse should feel free to leave because New Brunswick authorities would take care of the "spoilation" threat.

In his February 19 response, Fairfield remained defiant. He disputed Harvey's claim that the British had jurisdiction over the Disputed Territory until the boundary could be fixed. The governor was persuaded that "no such agreement . . . has ever been made between the two Governments. I have looked in vain for it among the numerous documents which have grown out of this question, and have never heard of any recognition of it, verbal or otherwise, on the part of the officers of the General Government." Moreover, even if there had been such an agreement, Fairfield held that it "can never be recognized by this State. A decent self-respect will ever forbid it." Since 1818, there had been an informal, unrecorded understanding between America and Britain that had allowed Britain to administer the Disputed Territory while the border issue was being resolved, but Fairfield was unmoved by this history.[16]

Further, the governor would not comply with Harvey's request that he recall the armed posse from Masardis because "the territory bordering upon the Aroostook River has always been, as I regard the facts, in the possession and under the jurisdiction of Massachusetts and Maine." He rested his position on the actions of the states over several decades, not on diplomatic agreements, noting that "that more than thirty years ago, Massachusetts surveyed and granted large tracts [in the Disputed Territory], which have ever been, in some way, possessed by the grantees." After Maine became an independent state, the rest of the territory "was surveyed and . . . divided between Massachusetts and Maine," and "both States have long been in the habit of granting permits to cut timber there without being molested from any quarter." In practice, the state had "exercised a jurisdiction which may fairly be regarded as exclusive over this territory."[17]

In Fairfield's view, it didn't matter if the boundary was still in flux. Massachusetts and then Maine possessed the territory because they had conducted business as if they had possessed it. Thus, when the governor learned that "a body of armed men [from New Brunswick] had gone into this territory and were cutting vast quantities of timber, defying the power of the State to prevent them," the state of Maine had sent an armed posse to the region. Near the end of his letter, Fairfield informed Harvey that "the party of the land agent now in the territory . . . will never leave it while the protection of the property of the State from plunderers renders it necessary for them to remain." If New Brunswick were to send a military force to expel the men, Maine would "do her duty" and "add to the glory of her career." The governor was certain that the citizens of Maine were willing to die for land, state pride, and white pines.

As a gesture of good will, Harvey released Rufus McIntire from Fredericton on February 18, and Governor Fairfield reciprocated by releasing James MacLauchlan from Bangor as soon as the land agent arrived in Augusta. But both sides had expressed irreconcilable views about who had jurisdiction over the Disputed Territory; and if everyone held to their positions, an armed conflict seemed to be inevitable. With this in mind, from mid-February into early March, both leaders took steps to put more men into the projected combat zones. On February 15, with McIntire still in custody, Fairfield had appointed Charles Jarvis as interim land agent. Jarvis had proceeded to the Masardis camp, where he was joined by hundreds of newly recruited civilian posse members. The governor also called out several divisions of the Maine militia, including hundreds of men who gathered in Bangor and then moved toward Masardis under the command of Isaac Hodsdon. Other frontline companies were sent to Houlton, located a few miles south of the Disputed Territory. Close to the New Brunswick border, the town was already home to a small U.S. Army garrison.

In New Brunswick, Harvey took advantage of his position as a British army general, which gave him the right to command existing regiments of regulars and summon militia units to protect the province from invasion. Given the importance of the Grand Communications Route, most contingents were sent to towns and settlements along this military road—among them Woodstock, the Tobique settlement, Grand Falls, and the Madawaska settlement. With the exception of Madawaska and garrisons along the northern section of the road, all of the deployments were within territory that everyone agreed was part of

New Brunswick. On both sides, the men had taken defensive positions, waiting to see what would happen next. But then, in late February, one small force made a move.

By February 17, Charles Jarvis had reached the Masardis camp. There he received a letter from New Brunswick's solicitor-general, Frederick Street, who was stationed at the confluence of the Aroostook and Saint John rivers. Street's message echoed Harvey's: the land agent was to immediately remove the Maine posse from the Disputed Territory. While the British commander was "very desirous to avoid any collision between Her Majesty's troops and any of the citizens of the United States that might lead to bloodshed," it would be necessary for the government to "take military possession of the territory" if the Maine men did not leave.[18] In a February 19 reply, Jarvis told Street that the "Solicitor-General of the provinces must have been misinformed as to the place where the force under my direction is now located, or he would have been spared the impropriety of addressing such a communication to me, a citizen of Maine." Didn't Street understand? Jarvis and his men were on Maine land, of course. The agent was willing to accept the possibility that Street's "impropriety" was due to an unfortunate error with respect to who owned the land. But if Street persisted in his "mistake" and attempted to remove the Maine men by force, then "deeply as I should regret a conflict between our respective countries, I shall consider the approach to [Masardis], by an armed force, as an act of hostility, which will be met by me to the best of my ability."[19]

After sending this letter to Street, Jarvis and his posse began to advance down the Aroostook River. By February 25, a few of his men had reached the area around today's Caribou, with no opposition from the British. The posse probably benefited from the diversion created when General Hodsdon and his Bangor militia changed their destination from Masardis to Houlton, a move that may have drawn the attention of British troops at Woodstock away from the Aroostook area. Seeing nothing in his way, Jarvis kept going. A day or so later, he and other members of his group stopped two or three miles from the undisputed western border of New Brunswick, not far from the Fitzherbert home where McIntire had been arrested. Jarvis and company settled in, building huts and fortifications and making plans to construct a boom across the river in the spring to stop the passage of trespass timber. The stronghold was dubbed Fort Fairfield, in honor of the governor; and on March 7, the newly released McIntire arrived to take command. Given the fort's location within sight of his recent capture, the moment must have been satisfying for the land agent.

In his communications with Governor Fairfield, Lieutenant-General Harvey had made his position clear: any armed American posse in the Aroostook River valley was to withdraw from the Disputed Territory. Yet not only had Jarvis and his men failed to withdraw, but they had also advanced to within a few miles of the undisputed border of New Brunswick and established a fort. New Brunswick militia and British regulars were already stationed within striking distance of Fort Fairfield, and it was Harvey's duty to order them to attack it and expel the Mainers by force. But he did not. Apparently, he believed that he had higher priorities: to protect New Brunswick from American invasion and to secure the Grand Communications Route where it passed through the Disputed Territory north of the Saint John River. Having already accomplished those tasks, he now chose to assume a defensive posture. In making this choice, Harvey may have been influenced by his time in the British army during the War of 1812. He had no interest in escalating into a third Anglo-American war but would rather sit tight and await further developments. In far off Washington, D.C., other events were unfolding during the last week of February, and they would confirm that Harvey had made a wise choice.

Usually, news traveled slowly in the 1830s; but as early as February 23, Henry Fox, the British ambassador to the United States, had received word from Harvey about the "unjustifiable incursion into a part of the disputed territory . . . by an armed body of Militia from the State of Maine." Wasting no time, Fox immediately wrote to U.S. Secretary of State John Forsyth, reminding him of the British view that the Disputed Territory was "under the exclusive jurisdiction of Her Majesty's authority." Therefore, "Her Majesty's officers . . . cannot permit any act of authority, such as is now attempted by the State of Maine, to be exercised within the territory in question."[20]

Forsyth replied on February 25, noting that Maine's actions had not been taken with military occupation in mind. Rather, "the sole object [had] been to remove trespassers who, in violation of the right of property . . . were gradually and hourly diminishing" the value of the state's land. Because New Brunswick had failed to protect Maine's white pines, the state had been forced to send in a posse. If McIntire had not been taken prisoner, the posse would have withdrawn on its own by mid-February—or so Forsyth claimed. The secretary of state's concluding paragraph must have come as a shock to Fox. He told the ambassador that the matter had been "laid before the President," who denied that Britain had exclusive right of jurisdiction in the Disputed Territory. Perhaps Van Buren had taken this position so he would be seen as standing up to the British, or possibly he really thought that the United States had never relinquished jurisdiction. Regardless, the diplomats needed to act fast to avoid war.[21]

On February 27, Ambassador Fox and Secretary of State Forsyth issued a joint memorandum intended to serve as short-term tool to reduce tensions. For the moment, they set aside the disagreement over jurisdiction; the plan was to address this later in a "friendly discussion" between the parties. In the interim, Her Majesty's officers in New Brunswick would "not seek to expel, by military force, the armed party which [had] been sent by Maine into the district bordering on the Aroostook River." However, "the Government of Maine [would], voluntarily and without needless delay, withdraw beyond the bounds of the disputed territory, any armed forces now within them." If there was a need to disperse "any notorious trespassers [that is, illegal pine harvesters] or to protect public property from depredation by armed force, the operation shall be conducted by concert, jointly or separately, according to agreement between the Governments of Maine and New Brunswick."[22]

In early March, Harvey received this memorandum from Fox, along with a letter making it clear that the British would not attempt to eject Americans from the Disputed Territory using military action. Harvey was pleased. He concluded that he must now defer "all offensive measures, as relates to the occupation by the Militia of the State of Maine, of a certain portion of the disputed territory." Instead of attacking Fort Fairfield, he would protect the Grand Communications Route "between this province and Lower Canada through the valley of the St. John and Her Majesty's subjects of the Madawaska Settlement [in the Disputed Territory]." In a letter to Fox, Harvey expressed an "anxious desire which I have always felt that matters of obviously secondary and minor import . . . should not be allowed to involve this province in [a] border collision with the State of Maine, which might lead to a national war." He may have taken satisfaction in the thought that his original choice to delay had averted a third Anglo-American war—at least for the moment.[23]

The Fox-Forsyth memorandum gave the nations a little breathing room, but the diplomats knew it might not hold for long. The British were more likely to remain patient. Not only was Harvey personally in favor of the temporary compromise, he was also working for the queen and would thus follow his government's lead. But Governor Fairfield was more of a free agent. He felt that he was answerable to the people of Maine more than to any American president. If Van Buren wanted to maintain the peace—and he did—then he would need to make sure that Fairfield was on board with any subsequent international agreements. To do this, he needed to send an envoy to Maine who was well versed in military and border matters, someone that Fairfield would have to respect

So Van Buren reached out to Major-General Winfield Scott.

As a veteran of the War of 1812, Scott had seen firsthand the mixed results of military action on the Canadian front. Early in the conflict, he had been captured by the British and, after being released in a prisoner exchange, had been wounded badly at the Battle of Lundy's Lane. Now, in 1839, he was serving as commanding general of the Eastern Department of the U.S. Army, which included maintaining peace on the Canadian-American border. Importantly, he was also well acquainted with Harvey. Though they had been enemies on the field during the war, they had since maintained a friendly relationship. For his part, Harvey knew that Scott was trustworthy: the American general had once returned a miniature portrait of Harvey's wife that had been found in a looted trunk after the American raid on York, Canada (present-day Toronto).

Though Scott expected Harvey to be easy to work with, he knew that Governor Fairfield and the Maine legislators needed to be handled with care. When he arrived in Augusta on about March 8, he "found a bad temper prevailing." It would be a challenge to craft an agreement acceptable to Maine that would prevent further confrontation over this "disputed territory," which Scott described as "a strip of land lying between the acknowledged boundaries, without any immediate value except for the fine ship-timber [white pine] in which it abounded." The general knew that "the only hope of pacification depended on his persuading the local belligerents." So he worked make himself personable. He took up residence at the hotel where many of the legislators stayed, and he "sat in the midst of them three times a day at the same public table." After riding with lawmakers "in a government sleigh" and discussing a potential agreement "in the vehicle," he attended a bipartisan supper in the town of Gardiner, a few miles south of Augusta. Scott thought the event in Gardiner went well, for "a feast is a great peacemaker—worth more than all the usual arts of diplomacy." In his memoirs, he was coy about exactly how he won the legislators' confidence, merely writing, "All the details of this negotiation cannot yet be given. There was, however, no bribery." It's notable, however, that he felt a need to explicitly state "no bribery"; perhaps this practice may was a standard operating procedure in such situations.[24]

By March 21, Scott had composed a general declaration, which Harvey signed on March 23 and Fairfield signed on March 25. It did not provide a permanent solution, but it would likely stabilize the situation until a treaty could be drafted. Through careful wording, Scott fudged the troubling question of jurisdiction in the Disputed Territory, recognizing instead that both Britain and Maine had occupied and administered portions of the land. The parties decided to avoid more haggling for the moment. They let de facto land possessions stand even as both sides retained the right to claim all of the territory. According to the signed statement, "Great Britain [holds], in fact, possession of a part of said territory, and the Government of Maine, [denies] her right to such possession; and the State of Maine [holds], in fact, possession of another portion of the same territory, to which her right is denied by Great Britain."[25]

Scott's declaration also addressed each side's primary concerns. Maine agreed that it would not "attempt to disturb by arms the said province [New Brunswick] in the possession of the Madawaska Settlement, or attempt to interrupt the usual communications between that province and Her Majesty's Upper Provinces" via the Grand Communication Route—a high priority for the Crown. In addition, "the Governor of Maine [would], without unnecessary delay, withdraw the Military force of the State from the said disputed territory." For his part, Harvey took pains to avoid any claim that Maine's militias were being driven out of the territory. He declared that it was no longer his intention "to seek to take military possession of that territory, or to seek, by military force, to expel . . . the troops of Maine." The militias would leave, but only because the state chose to withdraw them. At least, that's how it was intended to look. However, Maine's fury over illegal pine harvesting had not faded. Therefore, it wanted the right to deploy civilian companies to occupy and control key points south and east of the Saint John River to stop the cutting and transport of trespass timber. So the British agreed that Maine could leave behind, "under a land agent, a small [civilian] posse, armed or unarmed, to protect the timber recently cut, and to prevent further depredations."[26]

Scott's diplomacy ended the most dangerous phase of the conflict over the Disputed Territory. In essence, the Aroostook War was over. During the rest of March and into April, the Americans and the British pulled back their troops and sent the militias home. While the troops had suffered a few deaths from disease, accident, and winter exposure, no one on either side had been killed by hostile fire. It had been a splendid little war. Still, Maine was not convinced that the white pines were safe, and it took advantage of the clause allowing it to install a land agent to take control of the Aroostook River valley and its lumber. In the spring of 1839, Jarvis and his men constructed a sturdy boom of logs and iron links across the river at Fort Fairfield. Now they could stop and inspect any white pine logs being transported from forests in the river's expansive watershed. If the logs were determined to be trespass timber illegally cut by New Brunswick loggers, Maine authorities could seize and hold them until the loggers paid a fee or fine or until the pine was forfeited to the state. Fearing that New Brunswick lumbermen might destroy the boom, Jarvis proposed building blockhouses to protect the apparatus. By July, there was one blockhouse with an iron cannon at the site of the boom and a second on a hill about two hundred years away, protected by a palisade and armed with two artillery pieces, a twenty-four-pound howitzer, and a twelve-pound brass cannon.

In early September, William Parrott, an assistant land agent at Fort Fairfield, heard rumors that someone planned to harvest white pines in the valley using the false story that they carried timber-cutting permits from Massachusetts. Parrott responded by using the fort to block the passage of anyone with ox teams or other logging equipment. Lumbermen in New Brunswick were irate; and on September 7, between twenty and forty of them met at a familiar rallying point, Tibbets's store at the Tobique settlement. The men headed upriver, armed with muskets left at the store during the quarrels in the spring, and reached the fort at about two o'clock the next morning. But their surprise attack was foiled when a sentry spotted them and fired a shot in their direction. The New Brunswickers instantly retreated to Tobique. In keeping with the tradition of the Aroostook War, no one was harmed, and no other action was required.

Nonetheless, Warden James MacLauchlan was upset as he knew these events had the potential to reignite the conflict. He traveled from nearby Grand Falls to Fort Fairfield to meet with Parrott. The warden was accompanied by George William Featherstonhaugh, a geologist and geographer, who had been assigned to conduct yet another survey of the Saint John–Penobscot watershed boundary, which the British continued to insist were the highlands mentioned in the Treaty of Paris. In his journal, Featherstonhaugh called Parrott "a man of sense" who "considered the affair too ridiculous to be taken seriously." Although the New Brunswickers had gotten "within 30 yards of the pickets," they retreated like "men bereft of their senses, tumbling over the logs and stumps in the dark, and unable to find the road. Two muskets and five bayonets, three hats, an axe, some shoes and even boots were found [along the line of retreat]." In the end, nothing more came of the "raid" on the fort, and the peace held. Parrott "behaved with great kindness" toward the English survey party, probably guessing, correctly, that it would have little impact on the boundary question. Featherstonhaugh no doubt hoped his efforts would have much greater sway than they did. As he worked his way west along the British-favored highlands, he would occasionally carve "the name, or initials, of our Virgin Queen, V.R. [Victoria Regina]" on young beech trees to claim the land for England. Given that Queen Victoria married in February 1840, the first word of these carvings quickly became moot.[27]

Meanwhile, in another part of the Disputed Territory, Maine officials were focusing on the illegal cutting of white pines along the Fish River, which ran north into the Saint John. It was a ticklish proposition to put men at the mouth of the Fish to block the movement of trespass timber because this confluence was only twenty miles upstream from the British-controlled Madawaska settlement highlighted in the Scott agreement. According to the state's interpretation, the Scott agreement allowed Maine to install a timber-regulating posse at any location south of the Saint John River, including the length of the Fish. The British, however, asserted that the Madawaska settlement occupied both the northern and southern banks of the Saint John and extended as far upriver as the Fish. Because the settlement was an important point along the Grand Communications Route, they felt strongly that Mainers should stay far away from the mouth of the Madawaska River on the north side of the Saint John.

Nonetheless, in April 1839, the state sent an assistant land agent to the Fish River with instructions to build booms and blockhouses to stop the flow of illegally cut white pine. The agent, Alvin Nye, began by constructing a timber-blocking boom and blockhouse at Soldier Pond, about ten miles upstream (that is, south) of the mouth of the Fish. The British took no action, so he moved down to the confluence of the Fish and the Saint John and built a second boom and blockhouse combination. Nye named this post Fort Jarvis, after the land agent who had sent him to the region. (In 1841, the name was changed to Fort Kent to honor Edward Kent, who was then governor.) Now Nye could control the flow of white logs on the Fish River, but it also allowed him to expand his operations into the Saint John. Thus, when Warden MacLauchlan planned to drive logs seized on the upper Saint John past the mouth of the Fish, Nye claimed that the trespass timber belonged to Maine. In mid-May, Nye went so far as to lead an armed party of about thirty men up the Saint John to run off at bayonet point the river drivers whom the warden had hired to move the confiscated logs downriver. At this juncture, Harvey sent a letter to General Scott, complaining that Maine had violated the negotiated agreement by acting within the "Upper

Settlement of Madawaska" and declaring that he would take a diplomatic approach to the infringements. But Maine officials did not respond to Harvey's diplomacy, and they were no longer listening to anything that Scott might say. The state had staked out a claim on the southern bank of the Saint John, and they planned to hold it.

To solidify their claim that the Saint John's southern bank was American territory, the fort's occupants gathered on November 2, 1840, to cast votes in the presidential election. A British official, Francis Rice, was present at the meeting, and he objected to the voting and questioned its legality. The Americans duly noted his objections and then removed him, although Stover Rines, the new assistant land agent, intervened to limit any physical harm. Harvey considered what response he should make. He toyed with building a blockhouse on the northern bank of the Saint John, across from the Fish River's mouth, which would be garrisoned by an armed New Brunswick posse. In the end, the blockhouse wasn't built; but in December 1840, the British did send two companies of the Fifty-sixth Infantry Regiment to the Madawaska settlement to secure the military road. Then, in 1841, U.S. Army troops arrived to replace the civilian posses in both Fort Kent and Fort Fairfield. While this move probably reduced the chance of spontaneous and unplanned clashes between the two countries, it was also a sign that tensions were rising again. The nations needed to settle the border issue once and for all..

In late 1841, the new British prime minister, Sir Robert Peel, made it a priority to settle multiple British-American boundary questions, including the one that had aggravated and unsettled life in Maine and New Brunswick. He chose Alexander Baring, First Baron Ashburton, to represent the British in negotiations with the American secretary of state, Daniel Webster. Lord Ashburton was an appropriate choice. As a principal in the Baring Brothers banking empire, he had developed multiple ties in the United States. In the 1790s, even before his firm helped to finance the 1803 Louisiana Purchase, he had purchased vast tracts of timberlands in eastern Maine, when the district was still part of Massachusetts. He had acquired much of this acreage from William Bingham, a wealthy Philadelphian who at one point had owned 2 million acres (more than 3,000 square miles) of the district. Ashburton's woods were not in the Disputed Territory, but his acquisitions had given him a personal interest in Maine affairs. And his relationship with America hadn't stop with land purchases: he had married one of Bingham's daughters in 1798. Thus, with his Maine forests and his Bingham family connections, Ashburton would be open to compromise, and he was welcomed by the Americans involved in the negotiations.

The British strongly argued for drawing the border south of the American-favored line along the Saint Lawrence watershed. They did not want Americans in the highlands close to the river as this could allow them to block shipping in time of war. It was also essential for Britain to retain land north of the Saint John River to secure the Grand Communications Route and to keep at least some of the southern side of the Saint John River in the Madawaska region, not only because of that settlement's role in protecting the military but also because many of the inhabitants preferred to remain British. However, Ashburton and his government were open to compromising over most of the land south of the Saint John, including the Aroostook and Fish valleys. While many New Brunswick merchants and loggers wanted to keep these white pine lands in British hands, the acreage was expendable at the national level. By the 1840s, the British had other sources of white pines for masts; moreover, they sensed that the age of wind-driven warships was coming to an end. While the locals might complain, New Brunswick was still a colony subject to London's rule.

Secretary of State Webster had a greater challenge. While President John Tyler might not insist on retaining the entire 12,000 square miles of Disputed Territory, the citizens of Maine felt otherwise. That land and those trees were their inheritance, and the state wanted all of it, even up to the ridges overlooking the Saint Lawrence. While London could ignore New Brunswick's preferences, the United States was less able to ignore Maine's. In the years before the Civil War, federal authority was not nearly as strong as it would be afterward, and state leaders were liable to argue that the U.S. government could not give away a state's sovereign territory.

For a boundary agreement to pass muster, a significant portion of the Maine population would have to accept it. This meant that Webster would need to perform some deft political maneuvering during the negotiation process. He was helped by the fact that many Mainers were beginning to recognize that the British would not jeopardize the security of their military road without a bloody fight. As early as January 1841, Governor Kent was suggesting that "Maine, for the sake of peace and quiet of the country, and in her anxious desire to avoid collision with a foreign power, might forbear to enforce her extreme rights [to all of the Disputed Territory] pending negotiations."[28]

Lord Ashburton arrived in the United States in April 1842, and boundary discussions between Ashburton and Webster began in earnest in June, under the eye of Maine commissioners who had been tasked to watch them both. The American and British boundaries drawn in the 1820s were very far apart, and it had long been clear that neither side would accept the other's interpretation of the Treaty of Paris or the location of the highlands. To solve this problem, the participant would downplay the treaty's reference to highlands and draw a new border somewhere between the positions of the two nations. To a degree, a line had already been created by the establishment of British and American garrisons along the Saint John River and on its Fish, Madawaska, and Aroostook tributaries.

Briefly in July, British claims to the southern side of the Saint John River at the Madawaska settlement were a sticking point in the negotiations. But Ashburton eventually gave in, partly because the United States agreed that the western border of Maine would be drawn well to the east of the Saint Lawrence watershed boundary in those stretches where the highlands ran too close to British routes into Upper Canada. About eighty-five miles of the state's western border would be composed of two straight lines arranged at a slight angle to one another, drawn roughly parallel to and about thirty miles from the Saint Lawrence. Had the final border followed the highland, it would have been about fifteen miles from the river. This straight-line section would terminate at its southern end where the line intersected with the upper Saint John River; then the border would again follow the river until it reached its source in highlands separating the Saint John and Saint Lawrence watersheds. At this point, the watershed boundary was far enough from the Saint Lawrence itself that all agreed that the border could now follow the "highlands which divide those rivers that empty themselves into the River St. Lawrence from those which fall into the Atlantic Ocean to the . . . head of the Connecticut River." As for the northern border, while the citizens of Maine had sought possession of the land north of the Saint John for decades, few Americans had moved to this part of the Disputed Territory. The British would fight to hold onto the land traversed by the Grand Communications Route, and clearly it was not worth the cost or risk of mounting an invasion from Maine. So the northern border would follow the Saint John River, and the state would relinquish claims north of the river in exchange for fee-free transportation to the sea via the Saint John for all forest and agricultural products.[29]

Point by point, the participants were resolving disputes that had dragged on for decades. Perhaps progress was aided by the oppressive heat and humidity of a Washington, D.C., summer. No doubt the negotiators wanted to get the job done and escape from the mosquito-filled sauna on the Potomac. On July 1, Ashburton wrote plaintively to Webster:

> I must throw myself on your compassion to contrive somehow or other to get me released. I contrive to crawl about in these heats by day and pass my nights in sleepless fever.

In short, I shall positively not outlive this affair, if it is to be much prolonged. I had hoped that these gentleman from the northeast would be equally averse to this roasting. . . . Pray save me from these profound politicians.[30]

Anyone who has ever spent a summer in Washington can sympathize with him.

So the parties came to an agreement. Almost sixty years after the Treaty of Paris, the Maine boundary was drawn, and the Webster-Ashburton Treaty was signed on August 9, 1842. The British took uncontested control of about 5,000 of the 12,000 square miles of Disputed Territory, secured the Grand Communications Route, and reduced the risk that the Americans might block the Saint Lawrence River. Maine and the United States received 7,000 square miles and gained undisputed title to a vast acreage of conifer forest south of the Saint John River, including land where the government had created grids of townships, sold land and white pine cutting rights, and built booms and blockhouses to monitor the flow of timber. The question of who owned which white pines had been settled. And as we'll see, these thousands of square miles also contained other tree species, notably red spruce, which would prove to be just as valuable as the pines.

Chapter 4

White Pines and Red Spruce in the Nineteenth-Century Maine Woods

A LTHOUGH A BOUNDARY DISPUTE kept ownership of the white pines in northern Maine in limbo until the early 1840s, there were millions of acres of trees in the rest of the state; and since the end of the Revolutionary War, they had unquestionably been American. After the war, the population of Maine had risen rapidly, from about 100,000 in 1790 to almost 600,000 in 1850. As settlers migrated into the district, they cut down the forests to open the land for farming and turn the trees into cash. There were several valuable species in the woods, but the white pine remained the most commercially important during the first half of the nineteenth century. According to the Maine lumberman John Springer, "the amount of employment [that white pine] furnishes to lumbermen, mill-men, rafters, coasters, truckmen, merchants, and mechanics, exceeds that furnished by any other single product in Maine." Not coincidentally, the tree is featured in the center of state seal and is prominent on today's state flag.[1]

In colonial times, most of the harvested pines had been cut along the southwestern coast or close to the sandy banks of New England's lakes and rivers. But by the early 1800s, timber cruisers were obliged to expand into river lands located to the east and the north, where white pines were less common (see chapter 3).

They were also forced to search away from the rivers, deeper into old-growth forests, where it was harder locate the valuable trees among the other conifer and hardwood species. Young white pines need a minimum of 20 percent of full sunlight to survive; their seedlings and saplings struggle in the deep shade of mature forests. In very low light conditions, pines do not live as long as very shade-tolerant tree species do and may eventually be replaced by red spruce, hemlock, beech, or sugar maple. However, if wind or another disrupting force creates a light-filled gap in the conifer canopy, pine saplings can add as much as a foot and a half of height per year, shooting up at a faster rate than young spruce, fir, or hemlock. Rapid growth is essential to the white pine's survival: saplings in such gaps will reach a final height greater than that of the other tree species, which solves the problem of being shaded out. These tall trees then produce seeds and seedlings that may find new breaks in the canopy to begin the cycle again.

In Maine's interior old-growth forests, this pattern produced projecting islands of white pine within expansive seas of red spruce. For lumbermen in the early 1800s, the islands were the big prize; the ocean had little value. But how did pine hunters find their quarry in a vast and trackless wilderness? John Springer described their methods in his 1851 classic, *Forest Life and Forest Trees*. In autumn, a team of two or three men would travel up the Penobscot or Saint Croix River in skiffs or bateaus, armed with "necessary provisions" such as ship's bread, salt pork, tea, sugar or molasses, coffee pots, frying pans, woolen blankets for bedrolls, axes, guns, and ammunition. To spot the pines, they required a long-distance view, so the men would climb tall spruce trees, using the limbs as ladders. Because a spruce's lowest limbs might be forty feet from the ground, the pine hunters often cut down a smaller tree and leaned it against a taller one to gain access to the bottom branches of a lookout spruce. For an even better view, a "spruce-tree [might be] lodged against the trunk of some lofty Pine, up which we ascend to a height twice that of the surrounding forest."[2]

In Springer's telling, the woodsmen perched at the top of their lookout trees "like a mariner at the mast-head upon the 'look-out' for whales (for indeed the Pine is the whale of the forest)." They sought for "large 'clumps' and 'veins' of Pine." Springer wrote: "Such views fill the bosom of timber-hunters with an *intense interest*. They are the object of his search, his treasure, his *El Dorado*, and they are beheld with peculiar and thrilling emotions." To direct the men on the ground to the prize, a seeker at the top would point in the correct direction or, if he couldn't be seen from the ground, break off a limb and toss it in the correct direction. At the tree's base, another man would note the direction with a compass, a tool that was "quite as necessary in the wilderness as on the pathless ocean."[3]

Finding the pines was just the start. After locating a promising cluster, loggers had to select a site for their camp in the dense, unmapped forest. This would be the base of operations for the winter cutting season. A camp consisted of at least two buildings: a hovel that sheltered the ox teams and a cabin that housed the loggers. Springer described how the cabin was constructed. First, the men removed the top layer of leaves, turf, and litter, which might otherwise catch fire inside the cabin. Then they cut logs on site for the sides of the building, notching them so that the sides could be linked together. Spruce was their preferred building material because it was "light, straight, and quite free from sap." The roof was covered with pine, spruce, or cedar shingles, three or four feet in length, "not nailed on but secured in their place by laying a long heavy pole across each tier or course." Finally, the roof was "covered with the boughs of the Fir, Spruce, and Hemlock, so that when the snow falls upon the whole, the warmth of the camp is preserved in the coldest weather. The crevices between the logs constituting the walls are tightly calked with moss gathered from surrounding trees."[4]

The cabin had an open floor plan, with sections for eating, sleeping, and cooking "denoted by . . . small poles some six inches in diameter, laid upon the floor of the camp." Furnishings were spartan. In addition to a few stools, the cabin would contain a "deacon seat," described as "our sofa or settee." It ran parallel to the fire and was made from a spruce plank "some four inches thick by twelve inches wide, the length generally corresponding with the width of the bed, raised some eighteen inches above the foot-pole." Springer was mystified by the name of this piece of furniture: "It would puzzle the greatest lexicographer of the age to define the word, or give its etymology as applied to a seat, ... So far as I can discover from those most deeply learned in the antiquarianism of the logging swamp, it has nothing more to do with deacons, or deacons with it, than with the pope."[5]

The bed was communal so that body heat could be shared. "Our bedstead is mother earth, upon whose cool but maternal bosom we strew a thick coating of hemlock, cedar, and fir boughs. The width of this bed is determined by the number of occupants, varying from ten to twenty feet. Bed-clothes are suited to the width of the bed by sewing quilts and blankets together." The loggers' demanding work during the day allowed them to sleep deeply at night. "These hardy sons of the forest envy not those who roll on beds of down; their sleep is sound and invigorating; they need not court the gentle spell, turning from side to side, but quietly submitting, [they] sink into its profound depths."[6]

Once the camp was established, it was time to cut down trees. According to Springer, this was a team effort, requiring a boss in charge; choppers who selected, felled, and cut the logs; swampers who cut and cleared roads; a barker who stripped the bark from the part of the tree that would be dragged over the snow; and the goad or teamster who drove yokes of oxen. Carving a crude road out of the wilderness was a key step in the operation. This involved removing underbrush and cutting trees close to the ground to leave a path ten or twelve feet wide. The loggers did not have to create a smooth roadbed because "each fall of snow, when well-trodden, not unlike repeated coats of paint on a rough surface, serve[d] to cover up the unevenness of the bottom, which in time [became] very smooth and even." Springer took pride in these primitive routes: "We have here no 'turnpikes' nor rail-ways, but what is often more interesting. No pencilings can excel the graceful curves found in a main road as it winds along through the forest. . . . No street in all our cities is so beautifully studded with trees, whose spreading branches affectionately interlace, forming graceful archways above." Conditions would change in the spring, when "dissolving snows reveal[ed] . . . the stumps and knolls, skids and roots, with a full share of mud-sloughs, impassable to all except man, or animals untrammeled with the harness." But by then the loggers would be gone.[7]

The pines harvested deep in the Maine interior were not likely to be used for tall mainmasts. It was difficult to keep hundred-foot-long trunks intact and moving in the shallow rocky streams that led to the big rivers and then to the coast. Instead, before they began their journey downstream, the trunks were cut into thick logs, fifteen to thirty feet long. Already the work had been dangerous, and no doubt OSHA would not have approved of

LOG HAULING—PROCESS OF LOADING LOGS.

FIGURE 5. Oxen haul a white pine through the woods. From John Springer, *Forest Life and Forest Trees* (New York: Harper and Brothers, 1851).

their methods. But nothing equaled the perils of driving the logs through rapids, over waterfalls, and down streams and rivers to big log-shipping towns such as Bangor. Such work required skill and courage, and in the Penobscot watershed the men who took on the challenge became known as Bangor Tigers.

These risks were often multiplied during massive log jams, when thousands of trunks dammed the streams as immense pressure built up behind. According to Springer, unclogging a jam required "all the physical force, activity, and courage of the men, [most] especially those engaged at the dangerous points." One crew member might be "suspended by a rope round his body, and lowered near to the spot where the breach was to be made." He would search for a likely "key-log," whose extraction would break up the clog. Then he would fasten "a long rope to [the] log, the end of which is taken down stream by a portion of the crew, who are to give a long pull and strong pull when all is ready." The suspended man would pry at the key log while the men on shore would pull on the rope. Once the jam began to break, the man in the air would be "drawn suddenly up by those stationed above." This method could create its own set of hazards, for "in their excitement and apprehensions for his safety, this is frequently done with such haste as to subject him to bruises and scratches upon the sharp-pointed ledges or bushes in the way." Below him, "scores of logs, in wildest confusion, rush beneath his feet, while he yet dangles in air, above the rushing, tumbling mass. If that rope, on which life and hope hang thus suspended, should part, worn by the sharp point of some jutting rock, death, certain and quick, would be inevitable." When the jam broke, it emitted a "deafening noise, . . . produced by the concussion of moving logs whirled about like mere straws, the crash and breaking of some of the largest." This tumult, "together with the roar of waters, may be heard for miles; and nothing can exceed the enthusiasm of the river-drivers on such occasions, jumping, hurrahing, and yelling with joyous excitement."[8]

A river driver's agility and nerve couldn't always prevent

tragedy. Springer wrote: "No employment that I am aware of threatens the life and health more than river-driving." He noted that, when a river driver was killed, the others in the crew would improvise a coffin from two empty flour barrels fastened together: one covering the head, shoulders, and torso; the other the legs and feet. The victim was usually buried along the stream where he had died and "where, too, the lonely owl hoots his midnight requiem." A death affected everyone. "After such an occurrence, an air of sobriety pervades the company; jokes are dispensed with, the voice of song is hushed." When he returned in later years to visit some of these burial sites, Springer was

> oppressed with a feeling of sadness, while standing over the little mound which marked the resting-place of a river-driver on the banks of a lonely stream, far away from the hearth of his childhood and the permanent abodes of civilization. . . . [The] silent ripple of the now quiet stream (for the spring floods were past), the sighing of the winds among the branches of trees which waved in silence over the unconscious sleeper, rendered the position too painful for one predisposed to melancholy.[9]

Profitable white pine logging required a combination of people, vast forests, river transport, towns, sawmills, and a way to get products to distant markets. Beginning in the late eighteenth century, these components came together in spectacular fashion in south-central Maine. By about 1770, a handful of settlers had established a village on the banks of Kenduskeag Stream where it emptied into the Penobscot River. In 1791, this settlement was incorporated by the General Court (or legislature) of the Commonwealth of Massachusetts as the town of Bangor. The town was ideally located near the southern outlet of the massive, conifer-rich Penobscot River watershed, which covered 5,440,000 acres. Trees harvested in this watershed could be transported to Bangor by the river and its tributaries, and the flow of water could also power hundreds of sawmills. In addition, Bangor had a deep-water harbor that could be reached by ocean-going ships designed to carry away the output of forests and mills. The young town and the wilderness watershed were perfectly positioned to become a lucrative source of masts, lumber, and other wood products during the coming century.

But there was a lingering question. Who owned this land so rich in water courses and old-growth white pines? For centuries, the region had been the domain of the Penobscots, a subgroup within the Algonquin-language Wabanaki, who also included the Mi'kmaq of present-day Nova Scotia. Tribal members lived in both Penobscot River villages and seasonal camps as they followed the yearly cycle of fish and game across their vast territory. But in a pattern repeated throughout the Americas, the tribe had suffered a catastrophic decline in numbers when English and French explorers and settlers had arrived in the 1600s, bringing with them deadly European diseases. With their population so reduced, the Penobscots struggled to respond when white pine loggers and others colonial newcomers increasingly infringed on their territory in the late 1700s, mostly via travel up the Penobscot River valley through fishing grounds vital to the tribe's survival.[10,11]

In 1775, these intrusions prompted Chief Joseph Orono and other Penobscot leaders to travel to Watertown, Massachusetts, where rebellious American colonists had formed a provincial congress. (At the time, Boston was still in British hands.) Orono laid out the tribe's grievances to a committee appointed to confer with the Penobscots, noting that "[tribal] lands have been encroached upon by the *English*, who have for miles on end cut much of their good timber." The chief asked that the "*English* . . . interpose and prevent such encroachments for the future." In return, the Penobscots would assist the Americans in their struggle against the British. The appointed committee reacted favorably to Orono's request that the Massachusetts government take steps to "prevent any injuries" by outsiders and to his offer to form an alliance against the British. The soon-to-be-independent United States could use all of the Native allies it could get, for

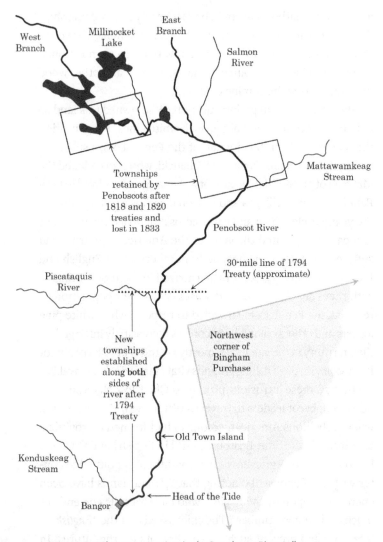

FIGURE 6. Penobscot treaty sites in the Penobscot River valley.

the British were actively recruiting tribes in other parts of North America to fight the American rebels.[12]

In a report to the provincial congress, the committee took the position that Massachusetts should "strictly forbid any person or persons whatsoever from trespassing or making waste upon the lands and territories or possessions beginning at the head of the tide on Penobscot River, [and] extending [north] six miles on each side of said river, now claimed by our brethren the *Indians* of the *Penobscot* Tribe." The phrase *head of the tide* referred to the point on the river where it was no longer affected by tidal fluctuations; this is located near today's Eddington, Maine, three or four miles upstream from Bangor. Crucially for the Penobscots, the twelve-mile-wide band of land running north from the head of the tide included a major tribal fishing ground at the Old Town falls, located at the southern tip of a large river island then called Old Town Island and known today as Indian Island. There was no mention in the report of who possessed the watershed lands beyond the twelve-mile ribbon centered on the river, but perhaps this was not considered significant at the time, given that the Penobscots felt free to hunt throughout the wilderness of central Maine. The provincial congress accepted the committee's report, confirming the treaty, known as the Watertown Resolve, and demonstrating that the Massachusetts government recognized that Orono's people held legal title to extensive tracts of land along the river. But would they be able to hold such titles in a new nation filled with ambitious men?[13]

After the Revolutionary War, an increasing number of American citizens, along with a few wealthy investors from Great Britain, were drawn to the potential of the Penobscot River region. At a time when most Maine roads were bad or nonexistent, a few acres along the river gave poorer settlers a place to harvest a few white pines for quick sale, land to clear for crops and livestock, and a watery highway for the transport of agricultural surplus to markets. Major land speculators hoped to make fortunes from the vast tracts of forested land in central and eastern Maine, including the territory around the lower Penobscot. These investors included General Henry Knox, George Washington's artillery chief during the war, as well as William Bingham and Lord Ashburton (see chapter 3).[14]

In the late 1780s, Bingham bought a million acres of wilderness in eastern Maine. His son-in-law, Lord Ashburton, later acquired a substantial portion of the Bingham Purchase, whose western edge was close but not adjacent to the Penobscot River. That's because a six-mile belt of land along either side of the river was owned by the Penobscot tribe. For Bingham and other investors, the tribe's title to this land was a barrier to their own economic success. The speculators needed it for sawmill sites, to control routes of transport from their forests to the river, and to increase the price of the interior lands they intended to sell to small-scale farmers, who would also want easy access to the river. As Ashburton explained, the land "would [be] difficult to settle with a wilderness of six miles between it and the finest river."[15]

So in the 1790s, major landholders pushed Massachusetts to procure the acreage along the Penobscot River. It was clear that valuable riverfront land could easily be sold to speculators, loggers, mill operators, and farmers, and the big landholders probably didn't have to work hard to persuade the government that acquiring this territory would generate revenue to pay off the Commonwealth's war debt. After considering the situation, state officials decided to negotiate a new treaty to transfer some of the Penobscot land to Massachusetts via a legal process known as title extinguishment. On February 26, 1796, the General Court resolved that "the Interest of the Commonwealth requires that the Claims of the Penobscot Tribe of Indians to certain Lands lying on each Side of the Penobscot River . . . from the Head of the Tide to the Source thereof should be ascertained and extinguished." This "shall be done by & with the free & voluntary consent of the said Indians—And the Principles of Justice, Humanity & Policy dictate that Compensation should be made for their releasing such Rights, & that some permanent annual Provision for their Support should be established by this Commonwealth."[16] The resolve spoke of consent, justice, and payment for territory, but clearly the legislators wanted those land titles extinguished, one way or another.

In the summer of 1796, Massachusetts treaty commissioners traveled to the new town of Bangor to negotiate with the Penobscots. Though he was now more than ninety years old, Chief Joseph Orono, along with several other chiefs, met them at the mouth of the Kenduskeag. By this time, population reductions had lowered the tribe's numbers to roughly 350 people, and the Penobscots were not in a strong bargaining position. When the discussions were over, the chiefs agreed to "grant, release, relinquish and quit claim . . . the Tribe's right, Interest, and claim to all the lands on both sides of the River Penobscot, beginning [at the head of the tide], and extending up the said River thirty miles on a direct line."[17] In other words, the treaty extinguished the title the Penobscots had held to a twelve-mile band of land stretching from today's Eddington north up the Penobscot to a point about two miles north of the river's confluence with the Piscataquis River. (The exact location of this line was not entirely clear.) In exchange, Massachusetts issued a one-time disbursement of blue woolen cloth for blankets, gunpowder and shot, corn, salt, hats, and rum, and it committed to a yearly delivery of Indian corn, powder and shot, and blue blanket cloth.

The treaty did not extinguish all Penobscot titles. Watershed land located north of the thirty-mile line still belonged to tribe; this territory included an important village located about sixty miles upstream of Bangor at the mouth of the Mattawamkeag River, along with vast expanses of white pine and red spruce forests. Critically, the treaty also excepted and reserved "to the said tribe, all the Islands in said River, above Old Town, including said Old Town Island, within the limits of the said thirty miles."[18] Old Town Island was about ten miles upriver from the head of the tide and so was well within the thirty-mile stretch of territory ceded to Massachusetts. But the Penobscots would retain their title to the river itself and to all the river's islands north of the southern tip of Old Town. Fishing grounds vitally important to the tribe's way of life would remain in their hands, which might potentially mitigate some of their losses in land.

But that hope would fade in the coming years as log drives, saw-mill dams, mill-associated pollution, and overfishing in the river below Bangor greatly reduced the number of salmon, shad, and alewives available to those on Old Town Island and further north.

After the treaty was finalized, the Commonwealth wasted no time in surveying the land that it had gained—nearly 200,000 acres. It divided the territory into about ten townships strung along both sides of the river, from Howland and Passadumkeag at the northern end to Bradley and Orono at the southern. The township of Orono was incorporated in 1806, five years after the death of Chief Joseph Orono, and one wonders what he would have thought of the choice of name for this territory so recently held by his tribe. Through land agents, Massachusetts sold the acreage in the new townships to small-scale farmers, sawmill operators, and the deep-pocketed investors and speculators who had lobbied for the 1796 treaty. What Ashburton had described as "a wilderness of six miles" between his properties and the river was no longer an obstacle to his prosperity.

In the years after the 1794 treaty, the Penobscots derived some income by leasing the right to white pines on the land north of the thirty-mile line. Although the number of sawmills downstream was still relatively low, this leasing system allowed the tribe to profit from the land they still controlled without needing to sell it. In 1801, it leased timber rights to loggers harvesting pines growing along the Mattawamkeag and Namadunkeeunk streams and, in 1802, to woodsmen working on land between Salmon Stream and the East Branch of the Penobscot, near present-day Medway. However, the Commonwealth of Massachusetts apparently had doubts about the tribe's ability to run its own affairs and in 1803 passed a resolution requiring Francis Goodwin, the state's super-intendent of Indian affairs, to approve all timber-related "bar-gains and contracts" before any lease could be considered valid. Goodwin would also "assist in the collecting of [the tribe's] just dues, and . . . prevent fraud and impositions being practiced upon them."[19] If he did not approve of a given lease, then it would be "utterly void," and any timber cut without the superintendent's consent would be forfeited. The violator would pay a fine triple the value of the timber, and money from the forfeiture would "enure to the use and benefit of the said Indians," although these funds would be controlled and distributed by Goodwin.

Under this paternalistic assertion of authority, the Penobscots continued to lease timber rights for cash and goods; and in August 1804, Joseph Treat, Josiah Brewer, and Isaac Hatch signed the largest leasing contract to date. This agreement rented the rights to harvest white pines along the Penobscot River from the 1796 treaty's thirty-mile line to Mattawamkeag Stream, located about twenty-five miles to the north. Pines could be cut for a distance of one mile on either side of the Penobscot and one mile on either side of tributaries up to a distance of six miles from their mouths on the main river. This expansive lease earned the Penobscots a total of $500, paid in five yearly installments. (For comparison, in 1800, a farm laborer might make about fifty cents a day while a carpenter could earn $1 a day).[20]

By the late 1810s, the tribe's population had dropped to roughly 250 people; and as more loggers poured into the upper reaches of the Penobscot River, tribal members found it increasingly difficult to prevent illegal or unleased timber cutting. Fishing had also become more difficult. So in 1818, the Penobscots asked the Massachusetts legislature to appoint commissioners to negotiate a new treaty so that the tribe could sell more of its land to the Commonwealth. The General Court was happy to oblige. They were eager to acquire more land in a territory that had become much more accessible and desirable. Not surprisingly, the treaty took away more land than the Penobscots had intended when they first petitioned the General Court. According to its terms, the tribe would "grant, sell, convey, release and quitclaim" all of their "right, title, interest and estate" to any and all land in the Penobscot watershed with the exception of Old Town Island, the islands upstream of Old Town, and four townships containing a total of about 100,000 acres.[21] Two of these were on opposites sides of the Penobscot at the mouth of Mattawamkeag Stream, and two were located on the West Branch of the Penobscot, to the west of today's town of Millinocket. In compensation for these hundreds of thousands of acres, the Commonwealth offered $400, calico cloth, ribbons, drums, fifes, knives, brass kettles, one cannon, and the promise of an annual October delivery to Old Town Island of various quantities of corn, wheat flour, pork, molasses, blankets, broadcloth (alternately red and blue), gunpower and shot, tobacco, chocolate, and fifty silver dollars.

In 1820, after the district split from Massachusetts to form its own state, Maine officials met with Penobscot representatives to sign a treaty that was little different from the 1818 treaty with Massachusetts.[22] But by the early 1830s, the Maine government wanted those four townships along the upper Penobscot River. As I discussed in chapter 3, the new state was grabbing as many townships as possible, eager for the sales revenue. Moreover, the township at the mouth of the Mattawamkeag was particularly desirable because it lay in the path of a military road that the state was building from Bangor to Houlton on the American-British frontier. So in 1833, the legislature sent two treaty commissioners, Judge Thomas Bartlett and Amos Roberts, to Old Town Island to negotiate an agreement for the sale of these four townships to the state. Roberts was a major player in the white pine business, but apparently the state was unconcerned about his obvious conflict of interest in negotiating a treaty that would bring him more profits.

The commissioners arrived at the island in the spring, negotiated the sale of the four townships for $50,000, and sealed the deal by acquiring the marks or signatures of fifteen members of the tribe, including Governor (or First Chief) John Attean and Lieutenant-Governor (or Second Chief) John Neptune.[23] At least that's how Bartlett and Roberts described the events in their June report to the state, which they drew up after their rapid return to Augusta. It was true that the commissioners had met with a few tribal members, but even some state legislators had doubts about their version of what had happened in Old Town. Significantly, the pair had not interacted with most of the tribe. Even if Attean and Neptune had fully understood the nature of the deal and had signed the treaty—and there are questions about these points—Penobscot governing and cultural practices did not give them the power to agree to a treaty without majority consent. And the majority was definitely not in favor of the deal described in the commissioners' report.[24]

The plan to sell the four townships created a furor among the Penobscots; and in the winter of 1833–34, Attean led a delegation to Augusta to request that the legislature void the June 1833 sale. They brought with them a written remonstrance arguing that the deed to the townships had been obtained by fraud and deception and that the sale did not have the consent of the majority.[25] In January 1834, the tribe's appeal was referred to the state senate's Committee on State Land for study and debate. The petition received some support but, in a fifteen-to-five decision, the committee voted to send a report to the senate stating that any legislation in favor of the Penobscots would be "inexpedient."[26] As a result, the legislature accepted the Bartlett-Roberts document as a valid treaty with a valid land sale, and the four townships passed into the hands of the Maine government. As for the $50,000 payment, that money would not go directly to the tribe but would be deposited in the state treasury and disbursed in annual payments "under the direction of the governor and council or said state, through the Indian agent for the benefit of said tribe; provided it should in their opinion, be required for the comfortable support of said tribe." After decades of negotiations and treaties, the Penobscots retained only Old Town and other river islands to the north—all that was left of a domain that had once measured in the millions of acres. As William D. Williamson, an early historian of Maine, wrote, "two centuries past, [the Penobscots] were the sole possessors of the country—numerous and powerful; now [this] tribe is driven to the islands of a river, once wholly theirs from its sources to the ocean."[27]

Meanwhile, loggers were pushing deeper into the Penobscot watershed in search of white pines, and sawmills were springing up on the banks of the lower river to process the logs floating downstream. By the 1830s, boom times had arrived in Bangor. As properties passed from one would-be timber baron to another, land prices, which had averaged fifteen cents per acre in the 1820s, spiked to as high as $10 per acre. In the 1860s, Hugh McCulloch, the secretary of the U.S. Treasury, described Bangor's heady atmosphere: "Broker's offices were opened . . . [and] were crowded from morning till night, and frequently far into the night, by buyers and sellers. All were jubilant because all, whether buyers or sellers, were getting rich." These were not experienced traders who had made a careful study of the timber business. In McCulloch's opinion, "not one in fifty knew anything about the lands he was buying, nor did he care to know as long as he could sell at a profit. . . . Buyers in the morning were sellers in the afternoon. The same lands were bought and sold over and over again, until lands which had been bought originally for a few cents per acre, were sold for half as many dollars." With speculation rampant, it was easy to take advantage of "lambs," as McCulloch called them. He described one scam in which buyers were deceived about the location of rivers needed to transport harvested tree: "Maps were prepared, on which lands were represented as lying upon water courses which were scores of miles away from them."[28]

Speculation was one thing, but turning promise into product was another. Dozens of mills were built in Bangor and in communities just upriver, including Old Town, Stillwater, and Veazie. To cut trees and transport them to the mills, the industry required thousands of forest workers, who would gather in rollicking Bangor at the start of the winter logging season and return in the spring to be paid for their labors. Predictably, businesses sprang up to meet the needs of these men, including numerous saloons and brothels. In his 1938 book about lumberjacking, Stewart Holbrook described the scene in Bangor, noting that undiluted rum sold for three cents a glass and that at least one bordello advertised its services for lonely woodsmen via the innocuous phrase "gentlemen's washing taken in."[29] All of this carousing was somewhat ironic, given that Bangor's name may have derived from a British hymn, but times had changed since the town's founding in 1791.

Assisted by the white pine, Bangor's population leaped to well over 10,000 inhabitants by the mid-1840s. However, it still felt like an outpost on the edge of a vast and ancient forest. In 1846, Henry David Thoreau described the town in his book *The Maine Woods,* calling it "a star on the edge of night, still hewing at the forests of which it is built." Though "already overflowing with the luxuries and refinement of Europe, . . . bear and deer are still found within its limits; and the moose, as he swims the Penobscot, is entangled amid its shipping, . . . and sixty miles above, the country is virtually unmapped and unexplored, and there still waves the virgin forest of the New World." As Thoreau traveled north from Bangor, he noted the loggers' preference for white pine: "The woods hereabouts abounded in beech, and yellow birch, of which last there were some very large specimens; also spruce, cedar, fir, and hemlock; but we saw only the stumps of the white pine here, some of them of great size. . . . It was the pine alone, chiefly the white pine, that had tempted any but the hunter to precede us on this route."[30]

The wilderness forests around Chamberlain Lake were a long way from Bangor, but they, too, contained an abundance of big pines and all of the water needed to drive logs to mills. There was one minor geographical problem: in the eyes of the Bangor timber barons, Chamberlain's waters flowed in the wrong direction. The lake's outlet on the northeastern shore fed a short stream leading to Eagle Lake, which was linked at its north end to Churchill Lake, which was drained by the north-flowing Allagash River. Thus, any logs leaving Chamberlain Lake by water had to bob north down the Allagash to the Saint John River, then to sawmills in British New Brunswick, and then finally to the Bay of Fundy and foreign markets. This meant that Bangor mills and ports would

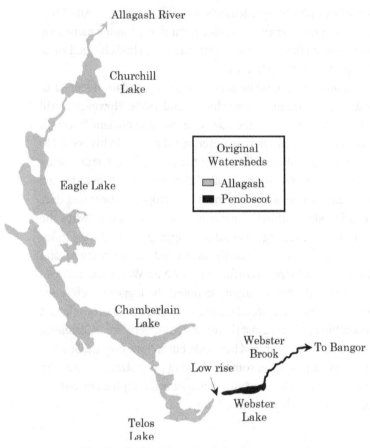

FIGURE 7. The Chamberlain Lake region before dams and canals.

miss out on the bonanza derived from harvesting, processing, selling, and shipping.

But geography could change. At Chamberlain's southern end, a short channel led to a smaller body of water called Telos Lake. It, too, was in the Allagash watershed, but its southern shoreline included an elongated tongue of water that created a cove pointing northeast. To the east of the cove was a low rise of land, about five feet in elevation above the lake. By looking over this low rise, one could see a dry gorge running for about a mile east to Webster Lake. Webster was about forty-five feet lower in elevation than Telos, suggesting that, in the past, Telos Lake may have emptied into Webster Lake. Webster Lake was in the Penobscot watershed: the lake fed Webster Brook, which flowed into the Penobscot's East Branch. In other words, all that separated the white pines around Chamberlain Lake from a water route to Bangor was a bump of land in a corner of Telos Lake. Nature seemed to be begging to float trees down the Penobscot, and the intrepid men of Bangor were determined to satisfy her.[31]

The plan was to build a low dam to block the northeastern outlet of Chamberlain Lake, let the lake level rise so its water would flow over the bump at the end of Telos Lake, and cut a canal in the gorge to Webster Lake. Then the bulk of Chamberlain's water would drain south instead of north. With the addition of a dam at the new outlet at the Telos Lake cove, timbermen could regulate the lake levels and the flow of water into the headwaters of the Penobscot River during log drives. In this way, the output from the Chamberlain Lake region could be transferred from the Allagash headwaters to the Penobscot basin. If asked, the Bangor men would have said that it should have been this way all along.

The plan to shift Chamberlain Lake into the Penobscot watershed began to take tangible shape in the late 1830s, when Amos Roberts (of Penobscot treaty infamy) and the brothers Samuel and Hastings Strickland bought a section of land designated as Township 6, Range 11 (T6R11) for $35,500 from the state of Maine. This purchase was a gamble, given that, until 1842, the British would claim to possess all of the Allagash watershed, including much of the land in this township. But the partners recognized that T6R11 contained key features for anyone who hoped to float pine logs to Bangor: a piece of Chamberlain Lake, nearly all of Telos Lake, the western half of Webster Lake, and the gorge connecting Telos and Webster. The three speculators began by cutting old-growth white pines around Webster Lake, with the short-term plan of using the lake to generate a pulse

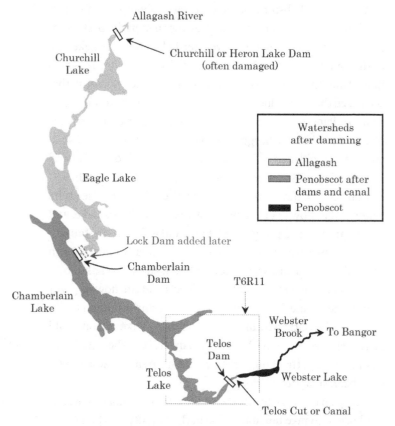

FIGURE 8. The Chamberlain Lake region with watersheds altered by dams and canals.

of water to flush logs down the main stem of the East Branch. But the pines of Telos and Chamberlain lakes were tantalizingly close, and they knew that, with a little engineering, logs could be driven through a cut between Telos and Webster.

Roberts and the Stricklands hired a man named Shepard Boody to work out how to move Chamberlain Lake water into the Penobscot. In 1841, he built the Telos Dam, a low gated barricade in the woods, about five hundred feet east of the Telos cove's shoreline, near the peak of the five-foot-high ridge that formed a boundary between the Allagash and the Penobscot watersheds. A second barrier, the Chamberlain Dam, was constructed at the northeastern outlet of Chamberlain Lake, which raised the surface of the lake by ten or eleven feet—enough to raise Telos Lake so that it now nearly reached the top of Telos Dam. The rising lake waters spilled into the passage leading to Webster Lake and the Penobscot watershed. When loggers cleared the ravine below the dam and excavated a channel, known as the Telos Cut or Telos Canal, the resulting rapid flow further scoured out the passage.

Operators at Telos Dam could release the water in pulses to drive logs toward Bangor. Just as importantly, the dam allowed Roberts and the Stricklands to count logs and collect tolls from loggers who had paid them for access to the white pines on their land. By the early 1840s, logs from the Chamberlain Lake region were passing through the dam and canal system and on to Bangor, and the landowners were making money. But Chamberlain Dam soon developed problems, and water levels dropped at the Telos Dam, slowing the flow of logs.

Enter David Pingree. Pingree was not a Bangor man but a Massachusetts shipping magnate known as "the Merchant Prince of Salem." During the 1830s, he had developed an interest in the timberlands of the Chamberlain Lake region. According to some accounts, this began when Pingree loaned money to a clerk in his office who had wanted to join the fun of speculating in the Maine woods; but when the clerk's endeavors failed, a chunk of Maine pinelands reverted to Pingree. His partner, Ebenezer Coe, also wanted to cash in on the Maine woods, but neither man was eager to slap blackflies in the Chamberlain Lake wilderness. Instead, they turned to Coe's son, Eben Smith Coe, a trained civil engineer, and sent him into the forest to study the prospect of turning trees into gold.

On his return, Coe reported that the land was filled with stands of white pine, many located near waterways. Convinced that he could make money, Pingree began to purchase huge tracts of land in the 1840s, eventually accumulating well over a million acres. Now that the 1842 Webster-Ashburton Treaty had settled the issue of national boundaries, his acquisitions were on firm legal ground; but before his investments could pay off, someone had to fix troublesome Chamberlain Dam. Back to Maine went young Coe, and by 1845 he had constructed an durable new dam. Once again, the waters of Chamberlain Lake rose to ten feet above their original elevation. Once again, logs could clear Telos Dam and shimmy down Telos Canal toward Bangor's mills and docks. Logging operators spent the winter of 1845–46 harvesting trees on land owned by Pingree and other speculators, and by the spring thaw they were ready to send their treasures downstream.

But David Pingree had made a mistake. As he and Coe were improving Chamberlain Dam, Amos Roberts was thinking about selling T6R11, the township in which Telos Dam and Telos Canal were located. He traveled to Massachusetts, planning to negotiate the sale to Pingree, but Pingree told Roberts to keep the land. Maybe he thought the asking price was too high, or perhaps he didn't want the hassle of operating a toll-collecting dam. Roberts increased the pressure, telling Pingree he would raise the toll to thirty-five cents per thousand board feet of pine. Pingree continued to pass on the chance to buy the land. Instead, he hinted that he would build his own canal from Chamberlain Lake, cutting through Mud Pond to Umbazooksus Lake on the West Branch of the Penobscot watershed, which would effectively render Telos Canal irrelevant. Negotiations failed, and Roberts returned to Maine. He doubted that Pingree would ever attempt this challenging engineering feat, and in fact no canal to the West Branch watershed was built.

Pingree had missed his chance, and his rationale is puzzling. He already controlled Chamberlain Dam and much of the region's timber. Yet his dam and landholdings would be far less valuable if there were trouble at Telos Canal because the cut was essential to moving his logs to mills and markets. By passing on an opportunity to buy T6R11, he was behaving like a Monopoly player who owns Boardwalk but then turns down a chance to buy Park Place, both of which he needs to win the game. Meanwhile, Roberts still had something valuable; and shortly after the unproductive meeting with Pingree, he sold the rights to Telos Dam and Telos Canal to Rufus Dwinel, a former mayor of Bangor. Dwinel owned timberland and sawmills, and he knew what to do with a monopoly.

In the spring of 1846, pine logs were backed up into the lakes like cars on the New Jersey Turnpike on the day before Thanksgiving, at least in the days before E-ZPass. But now Rufus Dwinel controlled the toll plaza. As the timber moved toward Telos Canal, he announced himself as the new owner of the cut: if you wanted to move logs through it, you would have to pay whatever he asked. Dwinel set the initial toll at thirty-five cents per thousand board feet, and timber operators howled. Rumor spread that they had plans to force their logs through the canal, so Dwinel hired up to seventy-five "distinguished citizens," many armed with long butcher knives, to supervise the cut. When logs arrived at the dam end of Telos Lake, loggers learned that the toll rate had skyrocketed to fifty cents per thousand board feet, due to the additional costs of paying the security force. In other words, the loggers would have to reimburse Dwinel for the expense of protecting the canal from themselves. But they had little choice; they paid the toll.

Monopoly triumphed, but Dwinel's victory was short-lived. That summer, a group of aggrieved landowners and logging operators petitioned the Maine legislature to solve the problem. The state may have taken an interest in the dustup because the canal affected the worth of the timberlands it served. A smoothly functioning canal would increase property values, the demand for land, and the size of the Maine timber industry, which ultimately meant more money in the state treasury. After several days of testimony, the legislature passed two acts to resolve what had become known as the Telos War. The first act allowed Dwinel to incorporate his Telos operations under a state charter, creating an entity called the Telos Canal Company.[32] The charter would permit him to charge a toll for logs passing through the canal. If he didn't receive payment within ten days of the timber's arrival in Bangor, he could put a lien on the logs and sell them at auction until his sales equaled the amount of the unpaid tolls. This was a win for Dwinel in that it allowed him to keep his chokehold on logs driven from Chamberlain and Telos lakes into Webster Brook and the East Branch of the Penobscot. However, the act also set the maximum toll that he could charge for moving logs through the cut at twenty cents per thousand board. Dwinel could no longer hike prices at will. Furthermore, if he did not accept the condition of this act by "the first day of October next," then "[it] shall have no further validity or effect." In that case, the legislators' second act would go into effect. It empowered his logging opponents to take over the Telos operations and incorporate them as the Lake Telos and Webster Pond Sluicing Company.[33] The toll would be set at ten cents per thousand board feet. Once the new company had earned enough money from this duty to cover its expenses, passage through the Telos Cut would be toll-free. In short, Dwinel had to either live with the first act's restrictions or give up the canal. He accepted the restrictions and the Telos Canal Company charter, ending the bloodless Telos War.

The original dams on Chamberlain and Telos lakes had been constructed to shift water and white pines from the Allagash headwaters into the Penobscot's East Branch, but the pines near these lakes would soon be logged out. Fortunately for Bangor's timber interests, there were many more pines around Eagle and Churchill Lakes, located to the east and north of Chamberlain. But they had the same geographic problem: they drained north into the Allagash River. Was there a way to send the region's pines down the Penobscot? This question drew the attention of Maine's government. The state still owned Eagle and Churchill timberlands and thus had pine-cutting rights to sell; and by sending this timber to Bangor instead of through Canada, they would keep more profits in Maine.

In 1843, Levi Bradley, the state's land agent, sent William Parrott to Eagle and Churchill Lakes to figure out a way to move Allagash watershed timber into the East Branch of the Penobscot. In his report, Parrott proposed two possible solutions.[34] Plan 1 involved building a tall dam on the Allagash River downstream of Churchill Lake. The dam would raise the surfaces of Churchill and Eagle Lakes by twelve and a half feet, putting them at the same level as Chamberlain and Telos lakes, if Chamberlain Dam were to be removed. The four lakes could then act as a single interconnected and continuous aquatic unit, all the way to the cove at Telos Lake. Here, a new canal could be dug through the five-foot rise between the Allagash and Penobscot watersheds to eliminate both the obstacle of the hill and the need for a dam on Chamberlain Lake. The plan appealed to Parrott because it required the building and maintenance of only two dams: the tall one at the outlet of Churchill Lake and a second one at the outlet of Telos Lake, which would control water levels and the entry of logs into Telos Canal.

Plan 2 offered a more complex solution, and it's not clear if this was originally Parrott's idea or was already being discussed among logging operators. As with plan 1, engineers would build a dam on the Allagash River downstream of Churchill lake, but it would be lower, raising the water levels of Churchill and Eagle Lakes by only eight and a half feet, about two feet below the original shoreline of Chamberlain Lake and about twelve feet below its surface, with a ten-foot-high dam at its northeastern outlet.

Why raise the level of Eagle Lake if its water couldn't top Chamberlain Dam? Were loggers supposed to flip massive white pine logs twelve feet over the dam and into the next lake? The idea appeared to be madness, yet there was method in it. The water-level rise in Churchill and Eagle Lakes would flood a short, shallow stream that connected Eagle Lake to Chamberlain Lake, allowing loggers to float timber from the two Allagash watershed lakes to near the base of Chamberlain Dam. The banks along the stream at the outlet of Chamberlain Lake were steep enough over a distance of about 1,500 feet that an elongated lock could be created by building a gated dam in this ravine. When gates in this lock dam were closed, any logs within the lock could be raised to the level of a dammed Chamberlain Lake by filling the cavity with Chamberlain water. When the gates in the original Chamberlain Dam were opened, the logs would float into Chamberlain Lake. Unlike plan 1, plan 2 would keep the waters of Churchill and Eagle Lakes in the Allagash watershed but would move timber harvested from those basins into the Penobscot watershed for driving to Bangor.

When Bradley presented Parrott's proposals to the legislature, he noted that, if the state were to build these proposed dams, it would incur additional and ongoing expenses related to operating

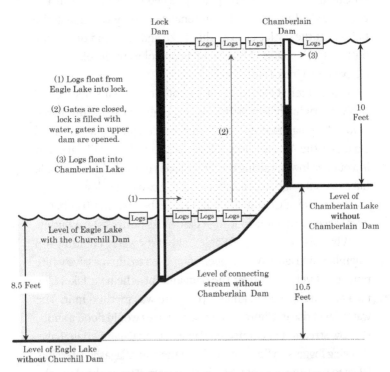

FIGURE 9. Dams and lock used to move logs from Eagle Lake (in the Allagash watershed) to Chamberlain Lake (in the Penobscot watershed).

and maintaining the system. Given those projected costs, he recommended that the venture be left to private companies.[35] It was time for David Pingree to step in again. Not content with converting pines to dollars at Telos and Chamberlain lakes, he had also invested in multiple townships adjacent to Eagle and Churchill Lakes. Like the state and like other logging operators, he preferred to send the harvest down the Penobscot, not the Allagash. Parrott had formulated two solutions, and Pingree chose plan 2.

In 1846, several parties interested in logging along Eagle and Churchill Lakes established the Heron Lake Dam Company. (At the time, Eagle Lake was called Heron Lake, so the company's name referred to that body of water, not to present-day Heron Lake, which is located at the north end of Churchill Lake.) The act incorporating this company authorized it "to construct and maintain a dam on the Allagash River at some suitable place below the Heron [Eagle] Lake so . . . logs and lumber may be transported therefrom, into Chamberlain Lake, and thence down the Penobscot River."[36] To turn the plan into reality, once again Eben Coe would be their man in the Maine woods, and that year he supervised the construction of new dams at the locations suggested in Parrott's 1843 report. One was built on the Allagash River at the outlet of today's Heron Lake, near the beginning of the Chase Rapids, and would become known as Churchill Dam, Heron Dam, or Chase Dam. The other barricade, a gated dam, was built on the south end of Eagle Lake between the high banks of the short stream linking Chamberlain and Eagle Lakes. This created a long lock between the existing Chamberlain Dam and the new lock dam. As a result of the lock's creation, at times, the original Chamberlain Dam was sometimes called the Lock Dam.

Though Coe and Pingree had invested considerable time, energy, and money in the new dams and lock, the system frequently did not operate at peak efficiency or at maximum capacity. By the spring of 1847, it was probably moving logs from Eagle Lake to Chamberlain Lake, but the Churchill Dam soon proved to be a weak link, susceptible to both natural and axe-wielding forces.

In the winter of 1847–48, crews led by Shepard Cary of Maine and John Glasier of New Brunswick were cutting white pines near Umsaskis Lake, the next lake downstream from Churchill on the Allagash River. When the lake thawed in the spring, river drivers moved the logs north down the Allagash and through Long Lake and Round Pond. Then the pines got hung up along the river below Long Pond, with the major obstacle of Allagash Falls yet to be crossed. Richard Hand, a member of the logging crew, later testified that a group of about twenty-five lumbermen armed with picks, hand spikes, and axes paddled back up the Allagash to the Churchill Dam. There, they tore a ninety-foot gash into the west side of the dam, and the resulting gush of water raised the river by several feet. As Hand put it, "the men got encouraged" by this reverse engineering, and the increased flow of water was enough to loosen the jam and drive the logs past Allagash Falls. They left the damaged dam for others to deal with.[37]

In the spring of 1849, Holman Cary and John Glasier paid a visit to the still intact Chamberlain Dam. As they approached, they noted that they were able to walk up the bed of the stream connecting Chamberlain and Eagle Lakes without getting their feet wet. This suggests that Churchill Dam was still damaged, as it had been designed, when fully functional, to flood the channel or ravine between the lakes. When Cary and Glasier later tried to move their winter harvest down the Allagash, the logs once again became jammed, this time between Long Lake and Round Pond. The Allagash men were convinced that Chamberlain Lake had more than enough water to serve those working on the Penobscot. Electing to bypass legal niceties, a logging crew traveled south to Chamberlain Dam, where they raised its gates, apparently without significantly harming the dam itself. The resulting large pulse of water flowed past the damaged Churchill Dam, and the Allagash drive was a smashing success.

It appears that the lock-and-dams system wasn't doing much better in the 1850s. Legislation in 1852 permitted the Heron Lake Dam Company to increase its tolls for logs passing through the company's dams and locks. However, it added the provision that the "company shall proceed . . . to repair and rebuild their said dam and works."[38] Apparently, the system was in need of a significant overhaul if it were to remain functional. A passage from Thoreau's *The Maine Woods* suggests that any repairs made in the early 1850s may have failed to solve the problems. In 1857, when the writer passed through the Chamberlain Dam area, he observed: "Below the last dam [the newer gated lock dam], the river being swift and shallow, though broad enough, we two walked about half a mile to lighten the canoe. . . . We were now fairly on the Allegash [*sic*] River. . . . After perhaps two miles of river, we entered Heron [Eagle] Lake."[39] Thoreau should have found a flooded channel at the lock's entrance capable of floating a log or an occupied canoe. Instead, he discovered only a shallow stream and then two miles of river between the lock and Eagle Lake. Apparently, Churchill Dam was not doing its job of raising the water level in Eagle and Churchill Lakes to the planned elevation, a key part of the plan to transport logs efficiently from the Allagash to the Penobscot watershed.

At the time, few observers considered or commented on the environmental impact of the timber barons' manipulation of the Upper Allagash watershed. But Thoreau saw and recorded the harm. While passing through Chamberlain Lake, he lamented the effects of the lake's higher water level on the original shoreline forest, describing it "as dreary and harborless a shore as you can conceive."

> For half a dozen rods [about a hundred feet] in width it was a perfect maze of submerged trees, all dead and bare and bleaching, some standing half their original height, others prostrate, and criss-across, above or beneath the surface, and mingled with them were loose trees and limbs and stumps. . . . Imagine the wharves of the largest city in the world, decayed, and the earth and planking washed away, leaving the spiles standing in loose order, but often of twice the ordinary height, and mingled with and beating against them the wreck of ten thousand navies.[40]

New legislation in 1860 addressed yet more trouble with Churchill Dam, this time due to flooding in 1858. Again, the legislature approved an increase in tolls, until such time as there was enough money to repair the dam, which had been "recently carried away."[41] In 1861, when naturalist Ezekiel Holmes passed through the region as a member of the Maine Scientific Survey team, he reported that Churchill Dam was "heavy and expensive . . . , having a lift of nearly 20 feet, and requiring strength to resist an enormous pressure of water, in order to flow back to the lock at the outlet of Chamberlain, not less than 30 miles."[42] But the dam was not in good shape and had "given away twice since it was built . . . and is not yet repaired." At the end of the 1860s, a hydrographic survey supervised by Walter Wells reported, "The lock [at Chamberlain Lake] is temporarily out of repair and is not used, [and] . . . the old dam at the foot of Churchill Lake . . . [is] mostly gone."[43] In the early 1880s, the writer and adventurer Lucius Hubbard published a series of tourist guides to Maine's northern lakes and rivers. In *Woods and Lakes of Maine: A Trip from Moosehead Lake to New Brunswick* (1883), he recorded that "the second or lower dam [at the Chamberlain site] is now in ruins. Its functions, those of the lower gate of a lock, ceased some years ago, at the time when Chase [Churchill] Dam at the foot of Churchill Lake was destroyed."[44]

The lock-and-dam network may not have functioned well for long, but probably most of the white pine logs for which it had been built were cut and transported before the system failed completely. Once the high-value pine was logged out, the woodsmen's attention turned to red spruce, and it may not have been worth the cost to repair and guard remote Churchill Dam. Even without it, timber harvesters could still send logs harvested along Eagle and Churchill Lakes down the Allagash River. Decades later, a new Churchill Dam would be built near the site of the old structure to again raise the level of Churchill and Eagle Lakes, but this time it was intended to regulate the flow of water down the Allagash. At Chamberlain Lake, the inoperative lock dam would continue to molder.

In contrast, the original Chamberlain Dam was consistently repaired and never abandoned. Unlike Churchill Dam, it was located close to Chamberlain Farm, a multipurpose lumber-industry outpost that Eben Coe had established in 1846 on the western shore of Chamberlain Lake. Large numbers of men were stationed at the farm, so they were easily available when Chamberlain Dam needed protection or repairs. The barricade remained essential to the operation of Telos Canal, which loggers continued to use to as a route to the East Branch of the Penobscot until about 1920. But the dam was always a sore point for disgruntled Allagash River timbermen. It had been designed to control the flow of Chamberlain Lake's water into either the Penobscot or to the Allagash. Yet even into the twentieth century, when the log drives ended, the lion's share of lake water continued to surge down the Penobscot, creating tensions between drivers on the two rivers. Now and again the Allagash crew would take extrajudicial action against the dam, and such exploits became part of Maine lore. For instance, they frame the plot of L. P. Wyman's 1927 novel *The Golden Boys Save the Chamberlain Dam*.[45] In that story, two teenagers, Bob and Jack Golden, thwart a plot to blow up the dam that has been concocted by Allagash woodsmen angry about the water flow. Though the book has long been forgotten, it is a cultural reflection of Mainers' long history of manipulating their lakes and rivers in service of moving white pine and red spruce.

These dams and canals had been built in the 1840s with white pines in mind, but times were already changing. White pine had never been abundant in the Maine interior; and by the mid-nineteenth century, most of the largest and most accessible trees had already been cut. By limiting itself exclusively to pine, the timber industry risked losing huge investments in hundreds of sawmills and millions of woodland acres. Fortunately, they were surrounded by the solution to their problem: the seas of red spruce circling the ever-smaller islands of white pine. Since colonial times, commercial logging operations had almost completely ignored the species. But now it looked like exactly what they needed to keep sawmills humming and profits soaring.

Red spruce is well adapted to Maine. The species thrives in cool climates with ample precipitation and can adjust to thin acidic soils such as those over granite bedrock or glacial till. In open areas, the seedlings do not compete well against fast-growing, light-loving red maples, birches, and aspens. However, they may survive at a higher rate in less fertile soils, where low nutrient levels inhibit the growth of hardwood competitors. In addition, spruce seedlings and saplings appreciate cool shade, which increases soil moisture, so the trees can develop for many years under the canopy of other species. In effect, less shade-tolerant trees put themselves out of business by helping to create a habitat in which the spruce can thrive. Eventually, the spruces outlive and outgrow much of their competition, just as the tortoise ultimately beats the hare.

In an undisturbed forest, red spruce trees can live for up to four hundred years and reach heights of one hundred feet with four-foot diameters near the base. The mature trees produce seedlings and saplings that grow slowly under low-light conditions, meaning that some understory spruce may be only five or ten feet tall after fifty years. Nonetheless, the shade created by the tall conifer canopy inhibits colonization by other tree species, allowing the spruce saplings to survive for decades. When gaps appear in the canopy as trees die or blow over, the enduring spruce saplings are in position to fill the holes in the forest. Thus, old-growth red spruce forests can perpetuate themselves for hundreds of years and may cover millions of acres in hospitable environments such as Maine's.

In the nineteenth century, red spruces were abundant in Maine, and now timber investors were noting that they also produced a soft, close-grained wood well suited for boards. The mature trees didn't approach the height and diameter of the magnificent white pines, but they were still large enough for conversion into lumber. Over the span of a few decades, the market for red spruce changed radically. Once it had simply been a tree that got in the way of pine hunters; now it was the state's dominant commercial species. Spruce was first harvested for mills and markets in the mid-1840s, but by 1860 the number of spruce logs driven down the Penobscot outnumbered pine logs. Twenty years later, spruce accounted for 70 to 80 percent of lumber production in Maine: the pine barons had become the spruce barons.[46]

The change in species created changes in logging methods. Most of the white pines harvested in nineteenth-century Maine had grown in separated stands relatively close to rivers and lakes. As I've discussed, woodsmen would build rough, narrow roads from small primitive logging camps to these isolated patches. But harvesting red spruce meant that loggers would be cutting down enormous ranks of trees, not one or two at a time. They needed a bigger and more permanent camp that could house much larger crews. A typical spruce camp included a bunkhouse (sometimes called a ram pasture) with beds along the walls, a combination cookshack and dining hall, blacksmith and saw-filer shops, an operations office, and a hovel for the horses that pulled the log-filled sleds. (Over time, horse teams had replaced oxen in woods work.) Although Bangor's Paul Bunyan statue features an ax-toting lumberjack, that representation is inaccurate. By the end of the 1800s, woodsmen felled most of the spruce with two-man saws.[47]

After swampers prepared an area by cutting a series of narrow skid roads, loggers would down the trees in nearly parallel rows alongside them. Then horse-drawn yard sleds would tow the timber from the stumps to skidways, or landings, located along larger routes known as "two-sled roads," where the logs were loaded onto rigs composed of two linked sleds. Horse teams hauled these double sleds to lake and riversides, and the logs were piled for driving in the spring. A contraption called a sprinkler eased the labor of moving many tons of wood per sled. This device consisted of a large wooden tank mounted on a sled and filled with hundreds of gallons of water. As a horse team pulled the sprinkler along a two-sled road, holes in the tank trickled water into the tracks. The water froze quickly, creating two slick traces for sled blades.

When Maine timbermen began harvesting red spruce in the 1840s, they must have felt as if the forests of large spruce trees would never end. But by the turn of the century, these trees, like the white pines before them, were rapidly disappearing. Not only was the supply of tall old-growth spruce for sawing into beams and boards disappearing, but Mainers were also facing rising competition from spruce operations in the Midwest. After reaching its peak lumber output in 1910, Maine's forest-products economy might have wheezed to a stop in the early twentieth century.[48] Instead, a technological change, dating back to the 1840s, saved the state's forest industry.

In the mid-1800s, inventors discovered a practical way to make paper using materials derived from wood fibers rather than pulped hemp, cotton, or flax fibers. When midcentury shortages led to increases in the price of rag-based paper, paper manufacturers began to switch over to the new method. First, fibers were extracted by grinding wood into pulp. The pulp was then treated with strong chemicals to dissolve the bits that were not fiber, and the remaining material was turned into paper. The process produced a skunky stench due to the various sulfur compounds involved, but it worked.

By the last decades of the century, a rapidly rising demand for wood pulp gave timber harvesters a new reason to cut down spruce. If the trees were to be used for pulp rather than lumber, short spruce trees were nearly as useful as tall ones. So woodsmen were now able to target smaller trees and saplings that they had formerly bypassed as unsuitable for longer boards. Regardless of their initial height, all trunks were cut into four-foot lengths for transport to the mills, which is why the pulp and paper industry continues to measure forest output in cords of wood rather than in linear board-feet, as the lumber industry does. (A cord is a woodpile constructed of four-foot-long logs and measuring eight feet long and four feet high, with a total volume of 128 cubic feet of wood.) By the late 1800s, a number of huge paper outfits had arrived in the Maine woods, including International Paper Company and Great Northern Paper Company. (At one point, Great Northern owned close to 10 percent of Maine's timberlands.) By about 1920, most of the state's spruce was being harvested for paper.[49]

As the twentieth century advanced, the timber industry made many changes in how it moved logs. Gradually, for instance, horse teams were replaced by mechanical log haulers. Patented in 1901 by Alvin Lombard of Waterville, Maine, these machines were first powered by steam and then by gasoline.[50] Their propulsion system was based on continuous bands of treads driven by sprocket wheels, making the machines look like small steam engines that had wandered off the tracks. With their caterpillar treads, they could ascend steep, snow-covered inclines that few other vehicles could climb. To guide the machine, a steersman perched on a platform at the front of a hauler and turned a wheel connected to a pair of skis below; steering was aided by icy grooves carved into snowy roads. Hauler crews often built shacklike shelters on the front platforms to protect steersmen from the subzero weather, but this could complicate a hasty exit in times of emergency. And emergencies were common. Lombard's design had dispensed with such fripperies as brakes, making the machine's downhill plunges exciting for everyone on the engine and the trailing sleds. Because the sleds would slide faster on the downslopes than the hauler did, jack-knifing was common, and everyone involved needed to be able to leap clear of a careening log train. Despite these dangers, and the haulers' tendency to break down in extreme cold, the machines could pull more than a dozen sleds at a time and were used in the industry until about 1930.[51]

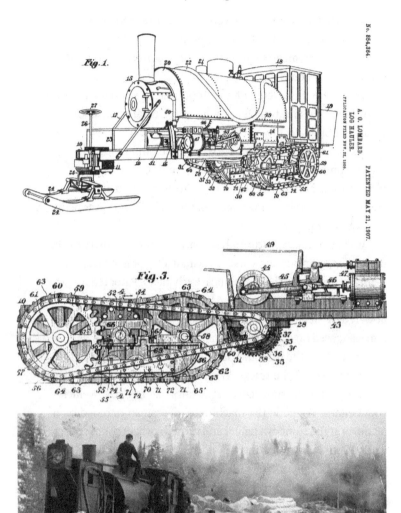

FIGURE 10. Lombard log hauler patent, showing tractor-tread system (top and middle) and a working hauler with a weather-screening shack on the front and sleds behind (bottom). The top two illustrations are from the U.S. Patent Office, 1901. The undated photograph on the bottom is from Digital Maine Repository, https://digitalmaine.com.

In the early 1900s, logging operators had not given up on the idea of moving timber from the Churchill and Eagle basins into the Penobscot watershed. Now they revived that dream using a method of log transport that would bypass the need for dams and locks. In 1902, the engineer Fred Dow constructed a tramway over the narrow divide between the western side of Eagle Lake and the northern end of Chamberlain Lake. The tramway first ran in 1903 and acted as a miniature railroad for spruce logs. Rather than being loaded into cars, the logs rode on saddles, toothed plates that were mounted on steel trucks or a wheeled chassis. The trucks were attached to a looping cable driven by a huge steam-powered sprocket wheel located on the Chamberlain Lake shore. The cable system pulled the logs across 3,000 feet of land and dropped them into Chamberlain Lake; then the empty trucks, still attached to the cable, would return to Eagle along a second track built beneath the log-bearing one. From the northern end of Chamberlain, a seventy-foot-long steamer would tow the loose logs south in enormous rafts contained by booms. From Telos Lake, the harvest would take the familiar route south through Telos Canal and down the East Branch of the Penobscot.[52] The tramway was a clever solution, and during its short heyday it moved about 100 million board-feet into Chamberlain Lake. But by 1907, it was replaced by another innovation: the Lombard steam log hauler. Timber operators saw little reason to haul away the tons of paraphernalia associated with the tram, and its cable, sprocket wheel, steam engine, and other relics still litter the land between Eagle and Chamberlain Lakes.

In the 1920s, Great Northern Paper Company financed the building of a logging railroad to shift the flow of harvested trees from one watershed to another. At the time the company was cutting red spruce for pulp in the Eagle Lake region, and managers decided that it would be cheaper to transport the trees down the West Branch of the Penobscot to its massive paper mill in Millinocket rather than move them via Chamberlain Lake, Telos Canal, and the twisting East Branch route. To reach the West

Branch drainage, spruce logs would be loaded onto railcars on the western shore of Eagle Lake and travel over a quarter-mile-long trestle built above a northern arm of Chamberlain Lake. The track would follow the edge of the lake for a few miles before turning southwest across the gentle rise separating the East and West Branch watersheds to the northern end of Umbazooksus Lake. From there, logs would float into Umbazooksus Stream, Chesuncook Lake, and the West Branch of the Penobscot. The plan echoed an old idea from the nineteenth century, when David Pingree and others had proposed digging a canal from Chamberlain Lake and Mud Pond to Umbazooksus Lake to move East Branch logs down a West Branch route, bypassing Telos Canal.

Nearly thirteen miles long, the Eagle Lake and West Branch Railroad was constructed by Edouard "King" Lacroix. Lacroix was part owner of the Madawaska Land Company, and he had long been accustomed to driving logs north into the Allagash River. Now he would be transporting logs south to the Penobscot tributaries. By August 1927, Lacroix's railroad was finished, and cords of red spruce were rocking down the rails to Umbazooksus Lake. The system operated around the clock in the summer and even had electric lights to illuminate the conveyor as it loaded logs at the Eagle Lake end. Ten or twelve cars, each carrying a little more than twelve cords, were pulled by two locomotives that themselves weighed nearly a hundred tons each. After a three-hour ride to Umbazooksus Lake, the cars were run out over the water onto a six-hundred-foot-long trestle. The bottoms of the cars were sloped, so when one side swung out on a hinge, the logs rolled out into the lake.[53]

The railroad was expected to operate for up to twenty years, but after only six years of use it was abandoned. By 1933, the nation's economic depression, combined with declining tree stocks, made the enterprise obsolete. The locomotives had originally been brought to the logging railroad in the dead of winter, when workers could take advantage of iced roads, a frozen

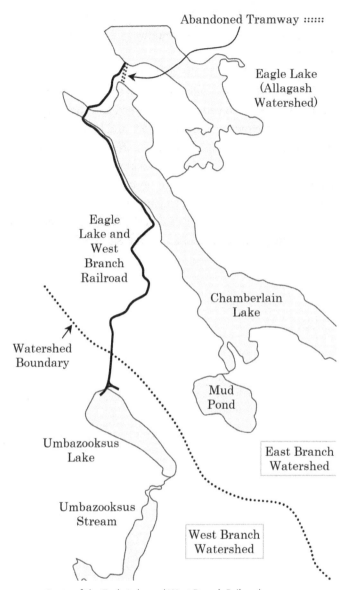

FIGURE 11. Route of the Eagle Lake and West Branch Railroad.

lake, and Lombard log haulers to slide the stock onto the new tracks. Now, in a replay of the tramway's end, they abandoned the engines in the woods.

The great challenge of the timber industry had always been the difficulty of moving bulky tree trunks to mills and markets. But as the twentieth century progressed and the state's road system became more extensive, paper and lumber companies began building their mills closer to the trees, and logging trucks became the preferred method of transport. The environmental damage of the river drives could no longer be tolerated. The old ways passed into history, leaving Fred Dow's tramway and Edouard Lacroix's locomotives quietly rusting away along the shores of Eagle and Chamberlain Lakes. *Sic transit gloria mundi.*

Chapter 5

Terpenes and Their Part in Christmas, Spruce Beer, and Tree Defenses

IN "THE WOODS OF Maine," published in *Harper's* in 1922, Dallas Lore Sharp described sleeping on a "mattress" of fir boughs, declaring that there is "no sleep [other] from which a man will waken half so fragrant and refresh[ed]." He longed to sleep on "this fir-balsam bed, for two or three weeks every summer, in the woods of Maine . . . [a] wish that . . . I coveted for city sleepers everywhere." The scents were "so clean and pure and prophylactic! They clear the clogged senses, and keep them in a kind of antiseptic bath, washing a coated tongue as no wine can wash it, and tingling along the most snarled of nerves, straightening, tempering, tuning them till the very heart is timed to the singing of the firs." Sharp "lay and breathed—as if taking a cure, this tent being the contagious ward of the great hospital, the Out-of-Doors." He wrote: "I breathed it to the bottom of my lungs; but my lungs were not deep enough; I must breathe it with hands and feet to get it all . . . and every vein ran redolent of the breath of the of the fir."[1]

Dallas Sharp was inhaling the Acadian Forest, and clearly he liked the scent. But what exactly was he smelling? In addition to providing us with many types of wood products, conifers are also chemistry labs that produce a wide range of fragrant substances known as terpenes. In addition to their pleasing aroma, many are very useful to us in other ways, and these molecules will be the focus of the next several conifer stories.

Terpenes are built from carbon atoms derived from carbon dioxide, which healthy green conifer needles capture from air during photosynthesis. Plant cells use these atoms to make chains of ten carbons. Then, much as a balloon artist turns a single long balloon into any number of animal shapes, conifers transform these carbon chains into an enormous variety of terpenes. They do this in one of three ways or in a combination of those ways: (1) bending the chains to make one or more rings of carbon; (2) adding or removing extra bonds between carbon atoms; and/or (3) adding oxygen atoms to produce terpenoids. If no additional carbons are added, then the final products are ten-carbon molecules called monoterpenes or monoterpenoids. In conifers, the most abundant monoterpenes include pinenes, limonene, phellandrene, carene, and bornyl acetate (a terpenoid), but these trees can produce hundreds of other related molecules as well. Species in dozens of other plant families also produce monoterpenes, and they are components of extracts known as aromatic *essential oils*. In addition to monoterpenes, conifers can also create fifteen-carbon chains that are then reshaped to produce fifteen-carbon sesquiterpenes such as farnesene, caryophyllene, and humulene.[2]

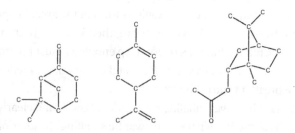

FIGURE 12. Examples of the carbon chains and rings of monoterpenes and monoterpenoids. (Hydrogen atoms attached to the carbon atoms have been omitted.) From left to right: beta-pinene, limonene, and bornyl acetate.

Walks in the Acadian Forest are enhanced by conifer aromas, which are composed of monoterpenes and sesquiterpenes. Monoterpenes evaporate or vaporize easily. When we inhale them, these volatile molecules stick to nerve cells in our noses, and we perceive them as a range of smells. Sesquiterpenes are also volatile and contribute to the trees' aroma, although they don't evaporate as easily and are usually present in lower concentrations. All conifers derive their scents from evaporating monoterpenes and sesquiterpenes, so why don't all of them smell exactly alike? The answer is that different types of terpene molecules have different scents, and genetic differences among species lead to variation in the types of terpenes produced and in the concentration of any given terpene. Different combinations of different amounts of terpenes produce different aromas for different trees. Firs, spruces, and pines all make carene, beta-pinene, and beta-phellandrene, but each species produces these molecules in different amounts so has its own unique smell. As in a cookie recipe, when you vary which ingredients are used and the amount of each ingredient, you get a different treat.[3]

Perhaps the best of these aromatic delights is the mix of terpenes found in balsam fir (*Abies balsamea*). As Dallas Sharp made clear, its aroma has long been an appealing and therapeutic part of the Acadian Forest experience. For many of us, the firs' wonderful scent is closely linked to Christmas, triggering memories of baking cookies, decorating trees, wrapping presents, and shopping for toys. Balsam fir is the classic Maine Christmas tree, and the industry makes millions of dollars from annual sales. Yet the industry began almost by accident.

By about 1880, the state had reached its agricultural peak, as measured by the number of farms, acres in cultivation, and amount of livestock. Ten years later, however, a survey requested by Samuel Matthews of the Commission of Industrial and Labor Statistics showed that 3,320 farms covering 255,000 acres had been abandoned or were unoccupied. By 1900, thousands of additional farms had been deserted. A large portion of this acreage was in Oxford, Franklin, and Somerset counties in the rough hills of interior western Maine.[4] The decline was hard to miss. In 1890, William Peter Atherton mourned in the *Maine Farmer*:

> The most noticeable thing on our way to Vassalboro [in Kennebec County] was the number of abandoned farms; we counted . . . [over] a distance of not much over six miles, eight farms large and small of that description, and a few others on the point of being abandoned. . . . Only ten years ago, one of these abandoned farms was valued at four thousand dollars [up to $200,000 today]; now, the buildings are rotting down, and the land fast growing up to bushes and weeds.

Atherton concluded with a benediction: "Peace to their ashes."[5]

In 1891, Joseph McIntyre, another writer for the *Maine Farmer*, described an abandoned farm in Wells, located in far southern Maine:

> [There was] an old cellar stoned up with round cobble stones, in which was a chimney partly fallen in decay. The house and barn had burned. In the front was a lot of old dead apple trees, where there had once been an orchard. . . . There was a clump of birch and stunted vines where there had once been a pasture, and a stretch of waste where there had once been a cultivated field. . . . A graveyard was in sight where the old settlers had been laid for their silent rest. The children (God help them) had fled to pastures new.[6]

Yet another *Maine Farmer correspondent*, Albert Pease, lamented in 1894, "Where have those people gone who used to derive a good subsistence from those farms? Where are the children whose happy shout and merry laughter used to ring through these now deserted 'banks and braes'?"[7]

The significant decline in Maine farming had many causes, and some weren't new. Though farming is never easy, stony, cold northern New England had always been one of the more difficult U.S. regions to cultivate. Throughout the nineteenth century, New England growers had been steadily moving to the Midwest and the Great Plains, where the soils were richer and deeper, with fewer stones and roots, where cattle could roam across vast rangelands. Maine farms did have the advantage of being relatively close to major eastern markets such as Boston, New York, and Philadelphia. Yet food was cheaper to produce in the Midwest, and railroad expansion made it profitable to ship products east. It was difficult for Maine farmers to compete with regions to the west in raising grains such as wheat and livestock such as beef cattle.

Moreover, the nature and methods of farming had changed over the course of the century. In the early decades, most farms were subsistence operations, growing crops for home consumption and producing their own milk, butter, cheese, eggs, poultry, beef, pork, fruits, and vegetables for home use. These homesteads brought in little cash, but that was not a major problem when expenses were low and limited to items such as clothing, tea, coffee, tobacco, sugar, and flour ground at a local mill. School taxes were low when teachers worked for about $1 a day in one-room schoolhouses; road taxes could be paid by labor in kind or by the loan of draft teams. Agricultural equipment was simple and inexpensive. The soil was broken with wooden plows; and, as an anonymous *Maine Farmer* correspondent wrote in 1891, "the stones and stumps did not materially interfere with the scythe, pitch fork and hand rake, which were the only tools of the merry hay maker of days gone by." According to this writer, "[the] cost of an outfit including team and implements to carry on a hundred-acre farm would not exceed one hundred dollars." Even death was reasonably priced, he declared, as there would be "a cheap coffin in the end."[8]

But in 1890, operating expenses for a modern farm were much higher. *Maine Farmer* writers pointed out that a farm now needed a "team of spirited horses attached to the sulky plow [a ride-on steel plow], the wheel harrow, the manure spreader, the planter, the horse hoe, mowing machines, tedder [used in hay harvesting],

and horse rake at a cost of . . . from five to seven hundred dollars."[9]
This up-to-date equipment also required "smooth and easily cul-
tivated fields," which was a problem for northern New England
farms, whose rocky, hilly acreage was often nutrient-depleted and
thus also required expensive fertilizers. Taxes were higher in this
decade because many citizens wanted better schools, and roads
and even the undertaker had additional fees. Farmers increasingly
faced a choice: find more cash from the sale of crops and animal
products or abandon their farms and move to the city.

Even if the parents chose to stay put, their children mostly
did not. Farms depended on a low-cost labor force, and until
now the unpaid children of big farm families had served in that
role. With the young people gone, farmers would have to pay
hired hands in cash and risk losing even more money. In 1891,
S. F. Emerson of Skowhegan, in Somerset County, explained in
the *Maine Farmer* that a "well-rounded crop of boys and girls"
was crucial to keeping fields in cultivation.[10] In his opinion, most
farms were abandoned because there were no children left to
take care of them. He wrote:

> [It is] sad to go by these homesteads and note the old cou-
> ple in their extreme lonesomeness, not knowing them-
> selves what is going to become of this home and the money
> they have labored so hard to accumulate. They have good
> attractive farms, and a splendid opportunity for the young
> couple, for the sons and daughters, but the young folks are
> not there. . . . [T]he outlook now is to soon step out, lock
> the doors, nail solid the windows, build a fence between
> the house and road and allow the grand old homestead, so
> associated with fond memories of other days, to be classed
> with the "abandoned."[11]

In 1891, a *Maine Farmer* correspondent writing under the
pseudonym "Aunt Patience" asked, "How shall we make farm life
and work so pleasant and agreeable to the rising generation that
they will esteem it an honor to follow the same calling of their
fathers and mothers instead of seeking for a living in some other

way?"[12] She suggested that parents should stop saying aloud "that farm work is about the hardest work there is, with such small returns." No wonder, argued Aunt Patience, "that the 'young folks' want to leave the old farm so soon." Maybe they would want to stay if the place were more like the farms of the past, when there were "good times at home, in the form of huskings and apple bees ending with unloading a well-loaded table of good things to eat; indulging in a few innocent games in which old and young participated." Perhaps, parents could "give young folks to feel if farm work is not the easiest, it is by far the most independent of occupations. Then there will be less abandoned farms and less desolate hearth stones." Aunt Patience's nostalgia is understandable, but huskings and apple bees can't hide the harsh realities of farm life. Whether or not the adults complained, the children understood that they would not be able to survive without change.

Less sentimental observers believed that some abandoned farms should never have existed in the first place. In the early 1800s, when Maine's population was expanding rapidly, newcomers were often forced to carve out homesteads in the stony hills because most of the better sites had been claimed already. At the time of purchase, many of these homesteads were still chiefly forest, so would-be farmers could survive for a while by cutting and selling the trees. But once the old growth was gone, the land often proved to be agriculturally unsustainable. According to a *Maine Farmer* report, a "speaker before a body of Connecticut farmers" noted:

> Most of the abandoned farms were left by their owners for good and sufficient reasons. Many of the farms were so sterile from the first that they were never worth carrying on, the men who cleared them having shown poor judgement in their selection. The first few crops took all the crop-bearing virtue the soil ever possessed out of it, and then there was nothing to do but to let the land grow up to bushes again.[13]

Maine Farmer writer William P. Atherton agreed that there was no profit in working "old, worn out, rocky farms," which produced goods that had to be transported "over hilly, horrible roads." He argued, "Such farms and such roads [had] answered their purpose, when time was of no account and man's wants were few and simple," but that time had passed for many marginal hill farms.[14]

Agriculture did not die out completely in Maine; the state would continue to produce farm products. Nonetheless, abandoned farms left hundreds of thousands of once-cultivated acres open to natural colonization. When cultivation or grazing stops, the ecological process of succession kicks in. First, a wide variety of shrubs and herbaceous plants takes over the habitat. Then tree species that do well in drier, sunnier, and initially less crowded habitats make an appearance and grow rapidly from seeds carried into the deserted fields by wind, birds, and mammals from nearby woodlots, tree lines, and isolated pasture trees. In the New England and Acadian Forest ecoregions, such trees include deciduous birches, aspens, and red maples along with conifers such as white pines and white spruces. These pioneering species create a little shade that keeps soils cooler and damper, and soon balsam fir seedlings begin to pop up. This species needs a little more shade and soil moisture than the earliest field-invading species, but before long that shade tolerance becomes a strength. Once established, the fir saplings compete well against the shade-intolerant colonizers as they slowly grow to an appropriate size for a living-room Christmas tree.

In the late 1800s and early 1900s, these fields filling with white spruce and balsam fir attracted the attention of speculators, who saw a market in bringing the sights and scents of a rural Christmas to cities far from Maine, New Hampshire, and Vermont. White spruce made a good-looking display, but its aroma was often compared to cat urine. Those trees might be best suited for outdoor displays. But many people loved the sight and smell of balsam fir inside their homes, and this was a lucky break for the landowners. In the past balsam fir had been considered a low-value tree and there had been little demand for the species, except locally at Christmas. But now wholesalers began prowling the backroads of New England in the autumn, hunting for the balsam firs that had grown up naturally in the old fields.

However, all of these lands were still private property. Someone held the deeds, and the wholesalers had to negotiate a price for the trees. Robert Frost's poem "Christmas Trees," published in 1916 in *Mountain Interval*, gives us an idea of how an exchange between a wholesaler and a landowner might have gone. Frost's poem is narrated in the first person and begins with the arrival of a "stranger to our yard" looking for something that former country people "could not do without and keep Christmas."[15] The wholesaler gets down to business: he wants the balsam firs growing in the woods, describing the forest as a place "where houses all are churches and have spires." But the landowner, who hadn't thought of his firs as Christmas trees, is reluctant "to sell them off their feet to go in cars / and leave the slope behind the house all bare / where the sun shines now no warmer than the moon."

Nonetheless, he agrees to walk with his visitor through the fir-filled former pastures, where some trees grow "in clumps [so] close that they lop each other of boughs," while others have a more pleasing form as they are "quite solitary and hav[e] equal boughs all round and round." After a tour, the buyer makes an offer: $30 for a thousand trees or three cents per tree. The landowner knows that his shapeliest firs would sell for up to $1 apiece in the city; there is no chance that he will sell to this man as "thirty dollars seemed so small beside the extent of pasture I should strip." But his time with the buyer has led him to view his balsam firs in a new light; and when he thinks of his friends in the city, he regrets that "I couldn't lay one in a letter." He concludes, "I can't help wishing I could send you one / In wishing you herewith a Merry Christmas."

Three cents per tree seems like an absurdly low price; but according to Thomas McAdam, writing in the December 1903 issue of *Country Life in America*, the offer was not unusual.[16] A "perfectly straight, symmetrical tree fully fifteen feet in height" might fetch forty cents from a buyer, but very few would meet that standard, and wholesalers were not roaming northern New England looking for a few prime specimens of balsam fir. Instead, they were purchasing most of the firs on a given acre and, within this random collection, the trees would vary considerably in quality, ranging from solitary beauties, to asymmetrical clusters, to contorted Charlie Brown specials. McAdam noted that, while some trees could retail for as high as $5, others might sell for only twenty cents. The wholesaler would absorb the cost of cutting, transporting, and distributing the trees to retailers. Among these expenses, moving the trees to market was usually the greatest. In the early part of the twentieth century, most firs were transported from Maine to East Coast cities by rail or ship, but by mid-century they usually went by truck on the expanding highway system.

Regardless of the costs, the business was very profitable for several decades. Writing near the beginning of this period, McAdam explained why the industry was likely to remain successful for many years. He wrote, "The best kind of Christmas tree is the balsam fir. Its needles persist longer than that of the black [or red] spruce. It is more symmetrical than the pine or arborvitae [northern white cedar] and it has strong, stiff branches capable of holding a good load of presents. This last consideration bars out the hemlock as one of the best Christmas trees." In a puzzling oversight, McAdam failed to mention the amazing aroma, but we can add that to its list of admirable traits. He also held that the "fir-tree industry seems to have no objectionable features. The balsam fir is too common, too widespread, and too eager to grow to be in danger of extermination." Once seen as nearly worthless, the trees now provided jobs to thousands of people in November, when most of the harvesting was done, and put "into the pockets of the people of the state . . . something like one hundred thousand dollars [per year]." Land that had become "worthless" was of value again. "It is said that one could sometimes buy a whole township for a hundred dollars. These same lands are now worth ten to fifteen dollars an acre for Christmas trees alone." Moreover, once the firs had been clear-cut, the land could be "restored to agriculture," if the landowner so desired.

During the early decades of the twentieth century, almost all of the balsam firs sold for Christmas trees had sprung up naturally, without plan or deliberation, in old fields and pastures. Farmers didn't have to take action to produce these trees, but they did learn certain tricks that could lead to greater returns. The trees were more valuable if they were symmetrically tapered, so some landowners thinned out the seedlings, creating space for the boughs to grow evenly around the trunk. Because wholesalers would pay less for trees that they had to clearcut themselves, some farmers and their families did cutting instead. They could also get a higher price if they wrapped the firs in bundles of three to six trees and transported them to central collection points. Sometimes they even sold them directly to consumers at farmstands, if local demand were high enough.

Into the 1930s, the market for fir trees continued to be strong. According to Austin Wilkins's report in a 1932 Maine Forest Service bulletin, "the Christmas-tree-cutting industry will undoubtably continue to be one of the most popular industries in Maine. Millions of trees will be sent each year by rail, water, truck, and parcel post to gladden the home on Christmas Day." Wilkins also offered a sneak preview of things to come: "A comparatively new idea is being introduced in Maine among landowners and tree growers, . . . the raising and managing of 'Christmas Tree Farms.'"[17]

But Wilkins's idea didn't take off right away. Twenty years later most of the holiday trees grown in northern New England were still being cut from old fields and abandoned farms. And business was brisk: Vermont, New Hampshire, and Maine each produced between 500,000 and 1 million Christmas trees in the early 1950s, with about half of Maine's trees coming from Washington and Hancock Counties in the eastern part of the state. However, the old field sources were becoming depleted. Most of the higher-quality trees had already been harvested, and stands that had been overlooked in earlier decades were now forests of firs that were too tall or misshapen for the Christmas market. Supply was not keeping up with postwar demand, and the number of Canadian imports was rising rapidly. Looking at the percentage of Boston's Christmas trees that had been cut in New England between 1941 and 1955, the Federal Reserve Bank's *New England Business Review* noted a decline from 98 to 60 percent, with similar declines in other eastern cities. Almost all of the gains in market share went to Canadian tree sellers.[18]

These declines in supply and market share prompted a shift toward tree farming. A balsam fir grown on a tree farm was far costlier to produce than one that wind or animals had planted naturally in an uncultivated field. But farmed trees could command a much higher price as they could be grown close to transportation routes and were more uniform in shape and quality. In addition, after harvesting a batch of suitably sized trees, growers could replace them with new saplings and thus guarantee a predictable supply. State extension office publications and state Christmas tree associations helped growers adjust to new methods, and today Christmas tree farms continue to do well in Maine and other parts of northern New England. But the industry got its start on "land fast growing up to bushes and weeds."

Conifer terpenes have powerful aromas, but they can also create powerful flavors. Much of what we think of as taste is, in fact, a product of smell—the stimulation of nerve cells in the nose as molecules in the mouth evaporate and waft up into the nasal cavity via a passage in the back of the throat. That's why we can taste terpenes when we add them to drinks or chewing gum.

In 1846, during an excursion to Mount Katahdin, Henry David Thoreau had the opportunity to sample a terpene-enriched drink: Maine-brewed spruce beer, which he was offered at the backwoods home of Thomas Fowler.

> Instead of water we got here a draught of beer, . . . clear and thin, but strong and stringent as the cedar-sap. It was as if we sucked at the very teats of Nature's pine-clad bosom in these parts,—the sap of all Millinocket botany commingled—the topmost, most fantastic, and spiciest sprays of the primitive wood, and whatever invigorating and stringent gum or essence it afforded steeped and dissolved in it,—a lumberer's drink, which would acclimate and naturalize a man at once,—which would make him see green, and, if he slept, dream that he heard the wind sough [whistle] among the pines. Here was a fife, praying to be played on, through which we breathed a few tuneful strains,—brought hither to tame wild beasts.[19]

Clearly, Thoreau found the drink to be a memorable one.

Spruce beer, known in French as *bière d'épinette*, is flavored with the buds, twigs, needles, or extracted oils of either black spruce (*Picea mariana*) or red spruce (*Picea rubens*). Descriptions of its taste and smell range from floral to citrusy to piney, all characteristic aromas of particular monoterpenes and sesquiterpenes. Most historical accounts refer to the use of black spruce for spruce beer, but it is easy to confuse that species with the closely related red spruce (see the appendix). The choice probably depended on a tree's relative abundance at a given location. Red spruce's range is confined to the northeastern United States and southeastern Canada. It is more drought-tolerant than black spruce; usually outcompetes in coarse, well-drained soils; and is more common in drier habitats and upland forests. The distribution of black spruce includes most of the red spruce's territory but also stretches far north to the Arctic and across Canada to Alaska. The species flourishes in the poorly drained soils of swamps and peat bogs; in these habitats, the trees may reach sixty feet in height and live for up to two hundred years. In Maine, where red spruce is more abundant, the beer was probably made most often with *Picea rubens* as the flavoring agent. It almost certainly was not brewed with white spruce, which has an unpleasant smell.

European accounts of brewing spruce beer date back at least to the 1700s, and the beverage may have originated with Native Americans, who sometimes used spruce foliage in unfermented teas. In about 1750, the Swedish naturalist and botanist Pehr Kalm traveled to North America and described a drink "used by Europeans in America . . . which is made from a species of spruce." He reported that the species was the same as the one listed in William Miller's 1754 *Gardeners Dictionary* as *Abies piceae foliis brevibus conis minimis.* Miller's common name for this species was "Newfoundland black spruce fir"—what we know as black spruce. According to Kalm, "the French in Canada are the foremost brewers of this small beer," in part because of the abundance of black spruce in the region. Beermakers extracted the spruce terpenes by boiling "branches and needles of the current year" in a kettle "until only half the water remain[ed]." Oily monoterpenes dissolve poorly in water; but by using heat, brewers were able to force the chemicals out of the foliage and into the water. Though many of the molecules escaped into steam, enough terpenes remained to give the beer its unique flavor. The French method included adding scorched grain or bread to the boiling kettle of spruce tips, which added "a golden-brown color to the beer" as well as a "pleasant taste" and made it "more nourishing." After cooling and filtering the liquid, "two or three stoups [tankards] of syrup [were] added. . . . Thus, the bitter taste which it ha[d] taken from the spruce twigs and resin [was] diminished." The syrup also served as a sugar source for the fermenting yeast that would convert the concoction into alcoholic spruce beer.[20]

Kalm's description reveals a key difference between brewing spruce beer and brewing ales and lagers. These traditional beers use barley or wheat as a source of carbohydrates for alcohol fermentation. However, it is difficult for yeasts to convert intact starch molecules into alcohol, so the process includes malting and mashing steps that break down the starches into sugars, which the yeasts then ferment. In contrast, spruce beer recipes use sugar syrups such as molasses as a source of directly fermentable carbohydrates. From the start, the carbs in molasses are almost entirely in the form of sugars, so malting and mashing are not required. As soon as the molasses is mixed with the spruce extract and other flavoring agents, the yeast can get to work, and the entire brewing process may take less than a day. Thus, spruce beer is much quicker and easier to prepare than traditional ales and lagers are.

In colonial Canada, even the wealthy drank spruce beer "because it is very healthful and quenches the thirst." According to Kalm, "most of the inhabitants, especially the French in Canada, have used the drink daily and find it does them a lot of good. Practically the only drink used by the officers and others in the forts is spruce beer." Kalm apparently believed in thoroughly immersing himself in the local culture because he "drank [spruce beer] often and found it very good. . . . It is of a clear brown color similar to near beer and has a good flavor. It tastes the least bit of turpentine and resin; however, this flavor is scarcely noticeable." [21]

By the 1800s, many people were promoting the idea that spruce beer was good medicine, and there were numerous available recipes. In *Elements of Materia Medica and Therapeutics* (1843), the pharmacologist and medical professor Jonathan Pereira offered a brewing method aimed more at pharmacists and physicians than at home brewers. Instead of using needles, buds, and twigs to make the beer, he suggested that pharmacists use *Essentia Abietis* (essence of spruce), an extract with high levels of terpenes produced by "boiling the young tops . . . of *Abies nigra* or Black Spruce . . . in water," followed by evaporation to concentrate the decoction.[22] This created "a thick liquid, having the colour and consistence of molasses, with a bitterish, acidulous, astringent taste." The pharmacist could then proceed to Pereira's recipe for *Cerevisia Abietis* (beer of spruce), which combined half a pint of the extract with flavoring agents such as bruised pimento, bruised ginger, and hops. This was mixed into three gallons of water and boiled for five to ten minutes, strained, and diluted with an additional eleven gallons of warm water. Then the brewer added six pints of molasses and a pint of yeast; and after twenty hours of fermentation, the beer was ready to drink. According to Pereira, you could imbibe it for pleasure or for health. The choice was yours.

As Pereira's directions demonstrate, nineteenth-century spruce beer recipes tended to call for extracts of spruce twigs and foliage rather than raw plant material. They also featured many flavor additions, including sassafras, wintergreen, ginger, hops, lemon, and pimento, perhaps in an effort to soften or disguise the turpentine taste noted by Kalm. Despite these innovations, spruce beer never achieved the popularity of traditional beer, and today it has almost disappeared from the marketplace. This suggests that beers flavored with spruce terpenes have limited appeal.

Monoterpenes and sesquiterpenes account for the aromas and flavors of conifers, but the trees can make even larger terpene molecules. They start with twenty-carbon chains and, after the usual modifications to create rings and new bonds between the carbons, the result is a diterpene or, if oxygen atoms are added, a diterpenoid. In conifers, most of the twenty-carbon diterpene-based molecules are diterpenoids such as abietic acid, which takes its name from the fir genus, *Abies*. As a rule, larger molecules with a few oxygen atoms (diterpenoids) evaporate more slowly than smaller molecules without oxygen atoms (monoterpenes). And instead of vaporizing easily upon contact with air, compounds such as abietic acid form sticky saps and resins that ooze out when the plant is damaged.[23] Humans have employed these conifer resins in numerous ways, as I will discuss in later chapters.

For now, let's look at an application that's fallen out of favor since it peaked in the nineteenth and early twentieth centuries: spruce resin as chewing gum.

In early September 1862, the newly organized Twentieth Maine Infantry Regiment was marching through the streets of Boston on its way to catch a train to the battlefields of Virginia, Maryland, and Pennsylvania. A wide-eyed private named Theodore Gerrish was absorbing these events and would later recount his experiences in *Army Life; A Private's Reminiscences of the Civil War*. As he recalled, the regiment was composed of close to 1,000 men, and "the sidewalks were covered with people who were eagerly looking at us":

> "Where are you from?" bawled an old salt, who stood leaning his back against a lamp-post. "From the land of spruce gum and buckwheat cakes," loudly responded a brawny back-woodsman from the forests of his native state. A loud laugh rang out from the crowd. One old gentleman swung his hat and proposed "three cheers for the old pine tree state."[24]

That confident backwoodsman knew his homeland well, for during the Civil War era Maine was famous as the land of spruce gum.

Spruce gum is made from resin exuded by a spruce tree when it has been injured or damaged by woodpeckers, squirrels, insects, extreme cold, or the loss of a branch. Some nineteenth-century spruce gum collectors, or "gummers," slashed V-shaped notches in the bark to mimic the natural damage that triggered resin flow. However, this approach required patience, for the tree might take up to five years after an artificial wounding to produce collectable quantities of spruce gum.[25] When it's released, spruce resin is whitish, soft, and less viscous than it will become; but as the monoterpenes and sesquiterpenes slowly evaporate, it hardens and turns into a dark, amber-colored gum with red or purple tints. This is the substance that gummers would seek.

It is not clear when or why people first began using spruce resin as chewing gum, but it was a familiar practice in nineteenth-century New England. Many books about life in the backwoods of the Acadian Forest include allusions to collecting and chewing it. Among them is a novel by Charles A. J. Farrar, who wrote several books in the late 1800s to draw tourists to the lakes and mountains of western Maine. (Not coincidentally, he had a considerable investment in the region's tourism industry.) In one of his publications for young readers, *Through the Wilds* (1892), he described four teenage boys discovering the mixed blessings of spruce gum:

> Ned and Dick had "struck it rich" on spruce gum, and their
> jaws were grinding like crushing machines. They each had
> a pocket-full and offered some to their friends. The Parson
> [a nickname for one of the boys] "took a chaw," but George
> told them, "it was too much like work, and didn't pay for
> the trouble."[26]

Spruce gum references even penetrated the rarified air of the *Harvard Lampoon*. In a satirical piece published on May 23, 1888, the president of the university's board of overseers reports the horrifying news that the "Freshman Base Ball Nine . . . has become addicted to the habitat of chewing gum" and declares that "it must be stopped at once."[27] One board member hesitantly points out that this would be "inconsistent with our great elective system where the men are allowed a right to their own chews." The president reluctantly agrees that compromise is required. To "stop the habit in some degree," the institution will limit the ballplayers to "chewing one particular branch of gum." Another board member recalls that his mother, back in 1763, "used to say that of all the gums, the gum of the spruce-tree was the best," and the president relates that he has been "informed by a prominent pitcher of the Freshman class that the whole secret of his success was the chewing of the spruce gum." Therefore, the one permitted gum will be . . . spruce gum! But lest students think that the board is allowing Harvard to descend into anarchy, "no man on the said nine, with the exception of the pitcher . . . [will] be allowed this privilege except on Saturdays and holidays."

When there is demand for a product, someone will step up to meet it. Most of the gummers were based in Maine, where red spruce trees grew plentifully. They usually did their collecting between October and April, when the air was cool enough to keep the resin hard and brittle and thus easier to scrape from the trees. No fancy equipment was required. Gummers used knives, long-handled chisels, hoe blades shaped into scoops, or gum spuds, a tool that Mary Rogers Miller described in *Outdoor Work*—a book intended to help young people earn money, build character, and become better citizens. Miller suggested hiring a tinsmith to shape a piece of galvanized iron into a funnel six inches deep, three inches wide at the top, and one inch wide at the bottom, with a two- or three-inch-long metal spout soldered onto the base. In her book, she promoted gumming as an activity for children because "a young man suffering from too little fresh air and attendant ills might find his health among the spruce trees while the gum paid the bills."[28] Gum pickers of all ages extended their reach by attaching their chosen scraping device to a long pole, and they collected the chipped resin in a bag hanging from the bottom of the scraper or by letting the gum fall onto a sheet spread around the base of the tree.

The gum they gathered varied significantly in quality and economic value. Lucky collectors might find clean, transparent, or translucent blobs of resin attached to a trunk, in amounts ranging from marble- to egg-sized. At the turn of the twentieth century, such high-grade gum might fetch $2 or $3 a pound, although prices did vary. These specimens were prized because they could be used directly as chewing gum with little or no additional processing; they were also in demand among druggists, who used the clean and nearly pure resin as a base ingredient in various pharmaceutical preparations. However, the bulk of most hauls was inferior-grade resin known as barrel gum or rough gum.

Because it had been scraped from the bark, it was contaminated with debris and would fetch only about 10 percent of the price of high-quality gum. Rough gum could not be used immediately but had to be heated until it was runny so that it could be filtered. While the processed material was still in a semiliquid state, gum makers would pour it into molds or spread it into cuttable sheets to produce pieces that could be individually wrapped for sale as chewing gum.

In 1932, Austin Wilkins reported that "there was a time when 150 tons of spruce gum valued at $300,000 [about $6 million today] were gathered annually in Maine," with the industry employing as many as "one hundred professional pickers and nine hundred occasional pickers."[29] But by the time he was writing, the annual spruce gum harvest in Maine had dropped to about five tons per year; and as chewing gum, spruce resin had been almost completely replaced by chicle-based products. Chicle,

FIGURE 13. Trade card advertisement for Martin's Spruce Gum and an advertisement for the Eastern Gum Company. From the Nineteenth-Century American Trade Cards Collection, c. 1870–1900, Boston Public Library; and *The New Rutherford Cookbook* (Wiscasset, ME: Emerson and Gray, 1916).

which is produced by Central American sapodilla trees (genus *Manikara*), is also terpene-derived. The inventor Thomas Adams introduced it as a chewing gum base in the 1870s, and eventually it came to dominate the industry. Chicle was easier to flavor, easier to chew, and could be gathered in much greater quantities. As a result, by the early 1900s, spruce gum had disappeared from most store shelves.

Perhaps the gum would not have faded from the market so rapidly if every package had included instructions for how to chew it. According to the *Maine Farmer*, a Vermont spruce gum dealer who was shipping the "luxury" to western markets in the 1880s found it essential to paste directions for use on each barrel: "How to chew spruce gum: 1st. Find a Yankee; he will show you how. 2d. If no Yankee is at hand, moisten or soften the gum in the mouth until it becomes soft enough to give to the teeth without crumbling. Never attempt to chew it when first placed in the mouth." Otherwise, the dealer said, the "western people" would "swear at the man they bought it of because he had sold them rosin, as they declared." The strategy must have worked, for "since the instructions have been sent, the western people make no more complaints." Directions or not, spruce gum remains an acquired taste, a challenge to chew, and probably not suitable for those with dental work.[30]

Why do plants work so hard to create a sticky and aromatic soup composed of molecules with rings and chains of carbons? The answer is that this biochemical bustling is a conifer's attempt to solve a fundamental problem. When damaged in a storm or attacked by herbivores, plants cannot run away. A mouse might flee from an owl, but when a spruce is faced with a ravenous moose, it cannot pull its roots out of the ground and scuttle away. Given that conifers may live for hundreds of years, they are bound to be found by herbivores many, many times. Somehow, then, they must defend themselves in place. For conifers, one way to do this is through the use of defensive chemicals such as the terpenes in their needles and under their bark. When injured, damaged, or attacked, conifers deploy their chemical counterattack in mixtures of oleoresins from an assortment of glands, blisters, ducts, and other terpene factories within the plant. The specific oleoresins and defensive structures vary from one species to another.[31]

The plant predators that encounter a conifer's defensive response are deterred by volatile monoterpenes and sesquiterpenes in the oleoresin that can repel or poison them. The larger diterpenes and diterpenoids may also be directly toxic to herbivores, but these twenty-carbon molecules also have an additional protective mechanism. When herbivore feeding exposes conifer resins to the air, the mono- and sesquiterpenes begin to evaporate, leaving the diterpenes and diterpenoids to form an increasing proportion of the terpenes in the resin. Before the damage, the smaller terpenes had functioned as solvents and had kept the resin runny and less viscous, but now air-caused changes in the ratio of terpene types make the resin thicker and stickier. In this way, the remaining larger terpenes act to glue together herbivore mouthparts, trap insects, and form a seal against further feeding as well as microbial infections.[32]

Terpene-based defenses are effective against most plant predators, but trees pay a price for them.[33] Carbon and energy allocated to defense are carbon and energy diverted from growth and reproduction, and the threat from herbivores can vary considerably over time. When the threat level is low, diverting resources to a high level of defense might be wasteful. So what's a tree to do? When should it distribute more resources to defense? The answer varies with the species, but one possibility is to maintain protective terpenes and resin ducts at a lower level until the tree is attacked. Then, when an herbivore begins to feed, the plant responds by ramping up defenses in a process called induction. The wounded or damaged plant tissue releases a signaling molecule called jasmonic acid that activates genes involved in terpene production and resin-duct formation. Depending on the species, the induction of defensive genes leads to (1) the production of more terpenes, (2) changes in the mix of terpene types, and/or (3) the construction of more ducts (called traumatic resin ducts) for the transport and distribution of the oleoresin.[34] It's worth noting that the induction of chemical defenses also occurs in many other plant species in a wide range of families and results in the production of many different chemical classes of toxic or repellant molecules.

Plants have evolved the ability to produce terpenes, and these chemicals discourage most herbivores most of the time. But herbivore species have responded by evolving adaptations to defeat plant defenses in a never-ending evolutionary arms race. For every conifer species, there is a collection of herbivores that can overcome that particular plant's unique mix of terpenes. Terpene tolerance or detoxification mechanisms vary with herbivore species, but tolerance is often a product of genes for enzymes that chemically alter the terpene molecules. Enzymes might degrade and break apart toxic terpenes, or they might alter the oily molecules to make them easier to dissolve in water. Increasing water solubility is a change that makes it easier to excrete the poisons in watery waste, and it's a mechanism that our bodies also use to protect *us* from toxins.[35] In some herbivores, the genes for detoxification enzymes are found in the microbes living in the animals' guts.[36] In these symbiotic relationships, the microbes provide protection against poisons, and the animals provide the microbes with a cozy home. In short, even with terpene defenses, every conifer species is vulnerable to some set of herbivore species somewhere. This becomes clear when we introduce non-native herbivores such as the red pine scale or hemlock wooly adelgid to the Acadian Forest, where the conifers have not had tens of thousands of years to evolve defenses against these specific threats (see the appendix).

Once an herbivore species develops the ability to tolerate defensive chemicals, it may continue to evolve to turn the terpene tables on the plant. Many herbivores that feed on conifers use a tree's unique mix of volatile terpenes to smell their way to the right species for feeding and egg laying. The terpenes may also reveal to the herbivore which particular trees are weaker and more vulnerable to attack. In some cases, insects even convert the consumed terpenes into volatile pheromones, or chemical messengers, to attract mates and invite other bugs to engage in a mass attack on the tree.[37] For certain pine beetles, pheromone-driven mass attack is the key to success because it exhausts the pine tree's ability to repel the invaders with terpenes. However, in one last plot twist, air-borne terpenes from both the trees and the herbivores may be sniffed out by insects that eat or parasitize those feeding on the conifers.[38] So terpenes may attract herbivores that specialize on conifers, but they may also send out a signal to predators and parasitoids that their dinner is ready. As Sweeney Todd says, "the history of the world, my sweet, is who gets eaten, and who gets to eat."[39]

In addition to inhibiting herbivores, conifer terpenes also have considerable antimicrobial activity. They can defend against disease-causing pathogens that often exploit openings created by herbivores, extreme cold, and storm damage. Many types of terpenes inhibit or kill the bacteria, fungi, and other microbes that would otherwise infect, damage, or destroy a tree. As Mary Rogers Miller wrote:

> Any injury of the living layer is like a "hurry call" to the cells where the resin is stored. These cells are the health department. They send out to the injured part a covering of balm, a salve which seals the wound. . . . We cannot say that the tree knows that the air if full of the germs of decay and that to let them get a foothold means decay and sure death; but the tree has something that serves the same purpose as knowledge.[40]

That is, in place of knowledge, trees have terpenes.

Conifers produce these antimicrobial terpenes for their own benefit, but molecules well suited for controlling microorganisms can also benefit us by preserving useful materials from microbial decay. For example, northern white cedar (*Thuja occidentalis*) is notable for its terpene-preserved, highly rot-resistant wood. Much of the wood's durability is due to high concentrations of thujone, a monoterpene with a slightly menthol-like aroma that takes its name from the white cedar's genus label. The timber is so long-lasting that you can still view the original Fort Kent blockhouse, built of cedar in 1839, at the confluence of the Fish and Saint John rivers (see chapter 3). White cedar is often used to make products that come into frequent contact with water or damp soil, including fence posts, shingles, buckets, barrels, tubs, and tanks. In Acadia National Park, it's the wood of choice for signposts on the carriage roads, at hiking trailheads, and on mountain peaks. Thanks to terpenes such as thujone, these posts will endure for years.

Chapter 6

Balsam Fir Resins in Medicine, Microscopes, and Germ Theory

ALEXANDER MACKENZIE (1764–1820) WAS not a doctor. He was a Scottish-born employee of the fur-trading North West Company tasked with establishing outposts in the Canadian interior and finding a water-based trading route from the interior to the Pacific Ocean. In 1778, Mackenzie and other company officials were sent from Montreal to Lake Athabasca to build a trading post called Fort Chipewyan in the northeast corner of today's Alberta. From there, in the late 1792, he followed the Peace River from Lake Athabasca to its western source as part of his search for a route the western ocean. On November 1, his party reached a fork in the river in what is now western Alberta and constructed the aptly named Fort Fork to serve as a base camp for the winter.

Mackenzie may not have been trained as a physician, but on that windswept riverbank in the Canadian prairie, he took responsibility for the health of anyone who asked if there were a doctor in the fort. Though he acknowledged his limited medical knowledge, "removed from all those ready aids which add so much to the comfort, . . . I was under the necessity of employing my judgment and experience in accessory circumstances, by no means connected with the habits of my life." In one case, one of his men "was attacked with a sudden pain near the first joint of his thumb, which disabled him from holding an axe. On examining his arm, I was astonished to find a narrow red stripe, about half an inch wide, from this thumb to his shoulder." The man was also experiencing severe pain, chills, shivering, delirium, and "several blotches on his body." It's likely he was suffering from a life-threatening case of blood poisoning, caused by streptococcal or staphylococcal bacteria that had entered a cut on his hand, but like everyone else at the time, Mackenzie knew nothing of bacterial pathogens. He ordered the man's arm to be rubbed with a liniment of rum and soap; when that failed, he followed the common practice of bleeding the patient, "perform[ing] the operation for the first time . . . from absolute necessity." Neither of these actions could have cured a blood infection, but the body has a remarkable ability to heal itself: the man survived and "regained his former health and activity."[1]

Mackenzie probably brought along a few remedies from Montreal (such as rum), but he also used local conifer species, including spruces and balsam fir, in his treatments. For example, in his journals, he described the case of a "young Indian [who] had lost the use of his right hand by the bursting of a gun." The man's thumb was left hanging by a small strip of flesh, and "his wound was in such an offensive state, and emitted such a putrid smell, that it required all the resolution that I possessed to examine it." Mackenzie was "alarmed at the difficulty of the case" but, given the desperate situation, "was determined to risk my surgical reputation." He stripped the bark from the roots of a "spruce-fir" (either a black or a white spruce), washed the wound in "juice of the bark," and then applied a poultice made from the bark. It proved "a very painful dressing," but soon the wound showed signs of improvement. At this point, the only connection between thumb and hand was flesh "shriveled almost to a thread," so Mackenzie removed the thumb. To complete the cure, he applied a salve "made of the Canadian balsam, wax, and tallow [animal fat] dropped from a burning candle into water." He continued to dress the wound three times a day for a month, and his treatment appeared to work. The wound healed, and near Christmas, Mackenzie's grateful patient brought him the tongue of an elk from a recent hunt and the "warmest acknowl-edgements" from the patient and his relations.[2]

We cannot know if the wounded hand would have healed on its own. However, if we assume that Mackenzie's treatment had a therapeutic effect, his salve's efficacy would have been a product of the terpenes in the Canada (or Canadian) balsam, the common name for the resin extracted from the blisters of the balsam fir. These terpenes may have aided healing by making the salve stickier and better able to adhere to the treatment site. And while doctors would not have understood this until the late 1800s, these molecules may have killed bacteria linked to skin diseases by causing damage to bacterial cell membranes. Laboratory studies suggest that balsam fir extracts can be fatal to the *Staphylococcus* bacteria responsible for many skin infections, although effects seen in test tubes do not always reflect accurately what happens when we apply antimicrobial terpenes to human bodies.[3]

"Doctor" Mackenzie's journal does not reveal if he learned about the use of Canada balsam from European colonists or from Native American sources. The indigenous peoples of inte-rior Canada had their own remedies for various illnesses, and Mackenzie was aware that they knew of the "medicine which con-sists in an experience of the healing virtues of herbs and plants." The healers of several tribes whose lands overlapped the balsam fir's range had concluded long before Mackenzie's journeys that the tree might have medicinal value. In one of many examples, the Potawatomi of the Western Great Lakes region drank teas made from the balsam fir's bark to treat lung complaints and used the liquid resin collected from trunk blisters when they had colds.[4] The terpenes would have had little effect on cold viruses, but the treatments may have provided relief from symptoms. They also used the resin in a salve to heal skin sores; as noted, fir terpenes have activity against skin-infection bacteria so may have had some medicinal benefits. Algonquin tribes far to the east, including the Wabanaki of northern New England, also used balsam fir resin as a skin antiseptic to heal sores, itches, insect bites, boils, and other skin ailments.[5]

Europeans arriving in North America observed the local tribes' medicinal use of balsam fir; and given their own centuries-long familiarity with the benefits of conifer resins and distilled oils, they were predisposed to adopt this species as a healing source. By the late 1700s and early 1800s, references to the medicinal use of balsam fir were appearing in American and European pharmaceutic dispensatories and materia medica. Such texts ran from several hundred to more than 1,000 pages in length and listed remedies for a wide range of ailments, along with directions for their preparation and application. Most treatments were plant-based; exceptions included alcohol, vinegar, and a mercury chloride salt called calomel. Calomel was known as the "Sampson of the Materia Medica" because of its alleged curative powers, but the use of mercury in medicine is a long and deadly story. Doctors and pharmacists used dispensatories and materia medica as guides for treating patients, and some published this information for a broader audience, with the goal of allowing anyone to act as their own doctor, regardless of formal training. For instance, John Gunn published *Gunn's New Domestic Physician* to provide medical knowledge based on the philosophy that the "fears of some that the physician should alone prescribe is a mistake."[6]

Medicinal substances derived from balsam fir were a standard part of nineteenth-century medicine, and they were prepared in a variety of ways. The foliage could be boiled in water to create a tea or decoction, the foliage or resin could be soaked in alcohol to make a tincture, or the resin could be used directly without alcohol extraction. As mentioned, balsam fir resin was usually called Canada balsam, and in *Elements of Materia Medica and Therapeutics*, John Kost described its source and means of collection: "[It] is obtained from the *Abies balsamea* in Canada and the State of Maine by puncturing the small vesicles which exist between the bark and wood of the trunks of the fir tree. As the [resin] runs out of the broken blisters, it is collected in bottles."[7] John Gunn noted that balsam resin "is an article found in all the drug-stores" and that anyone could use it without prescription. To help readers identify the genuine article, he described it as "the juicy or resinous exudation" of the balsam fir tree and "a semi-transparent fluid, nearly colorless, or of a slightly yellowish tint, about the consistency of honey or thick molasses, [with] a slightly bitter taste, and of rather an agreeable odor."[8]

Some nineteenth-century manuals described treatments that explicitly called for balsam fir extracts and resins. Others grouped Canada balsam with the resins of other conifer species and then discussed the medical applications of all of them together, without making many species-specific distinctions. However, while pharmaceutical texts often discussed the saps of different conifer species as if they were interchangeable, most commentators believed that Canada balsam was superior. William Lewis called the fir extract *Balsamum Canadense*, describing it as "one of the purest of the [resins]" and noting its "very agreeable smell, and warm, pungent taste."[9] In contrast, he considered the resin of common pines to be inferior, coarse, heavy, and "in taste and smell the most disagreeable of all the sorts."[10] Still, pine resin was available in greater quantities and at lower prices than balsam fir resin was, and it was probably used much more often. Pine sap was also distilled to produce what was called oil of turpentine to distinguish it from raw pine sap or resin (see Chapter 7).

Dispensatories and materia medica offered many fir-based remedies and treatments, with the American and European texts paralleling the Native American use of fir resin for skin afflictions. Doctors, pharmacists, and amateur practitioners combined the resin with other ingredients to generate salves and plasters that they used to treat burns, wounds, and skin infections. In *The American Vegetable Practice, or A New and Improved Guide to Health*, Morris Mattson claimed that a balsam fir plaster "will heal bad wounds in a very short time [and] it is also applied to burns and scald with great benefit."[11] As I've mentioned, such treatments might have had beneficial effects, due to the terpenes' antimicrobial properties. However, it's worth noting that our cells also have cell membranes vulnerable to terpene damage, and these molecules may be as toxic to human cells as they are to disease-causing bacterial cells. Still, as long as practitioners applied antiseptic conifer extracts onto skin instead of administering them internally, patients may have been able to tolerate the terpenes. In such cases, the fir resin was in direct contact with the bacteria to be eliminated, and the skin itself acted as a barrier against toxic chemicals. Skin's outermost layer is composed of cells that are already dead; and while some terpenes can pass through skin, the absorption of toxic molecules across the skin and into the blood and deeper tissue layers is less than when toxins are absorbed across the surface of the gut or lungs.

The situation changes when antimicrobial chemicals must reach pathogens causing infections inside the body. When taken internally, a dose of balsam fir terpenes high enough to have a therapeutic effect might also cause significant injury to the cells of internal organs. A smaller dose can reduce damage, and humans do have the ability to detoxify and eliminate terpenes via urine and by evaporation from the surface of our lungs. Yet these factors also reduce the probability that the treatment will produce a cure. This dilemma is not unique to terpenes. Healers long struggled to identify chemicals that could be used at doses that would kill dangerous internal bacteria while causing relatively little harm—a problem that was solved only in the twentieth century, with the discovery of penicillin and other antibiotics. These miracle drugs target aspects of bacterial cell structure and physiology that are unique to, or significantly different from, the microbes'. Therefore, they are much more toxic to microbes than to us, allowing us to use these drugs internally at concentrations we can tolerate.

The potential poisoning problem did not stop nineteenth-century healers from treating internal ailments with oral medications containing Canada balsam. At that time no one was running clinical trials to determine if a given remedy were ineffective or too toxic. Nonetheless, dispensatories and materia medica freely recommended oral fir resin preparations for coughs, chest congestion, asthma, laryngitis, bronchitis, "pains in the breast, and incipient [tuberculosis]."[12] They also promoted fir-derived treatments for urinary- and reproductive-tract infections. Many texts noted that remedies made from balsam fir had a diuretic effect; and in *Elements of Materia Medica and Therapeutics*, John Kost confirmed that Canada balsam was "used in the treatment of [diseases] of the urinary organs, particularly those characterized by mucous discharges," including sexually transmitted diseases.[13] Depending on the specific medical condition, doctors would prescribe mixtures of resin and sugar, pills rolled from resin and other thickening agents, emulsions combining resin with egg yolks, or electuaries (pastes) of resin and honey.[14] Most treatments delivered the equivalent of one to three grams of balsam fir resin per dose, which would be taken one to three times per day. It is unlikely that any of them were useful in curing respiratory, urinary, or reproductive-tract afflictions.

Intestinal complaints were also treated with balsam fir remedies. When combined with sugar, doses of Canada balsam or resin-derived tinctures were prescribed for mucous diarrhea, bowel ulceration, stomach soreness, and hemorrhoids. In the 1841 edition of *The American Vegetable Practice*, Mattson asserted that "a decoction of the bark is an excellent remedy in diarrhea and dysentery." His was made by boiling shredded fir bark for half an hour, straining it to remove the debris, and sweetening it with sugar. The resulting viscous liquid had a balsamy, slightly bitter taste, but Mattson assured parents that "children will take it freely, and it is particularly valuable in the bowel complaints with which they are so often affected." He promised, "It rarely fails to cure, even in the very worst cases. It diminishes the pain and soreness of the bowels, and gradually checks the discharges."[15] If Mattson's claims had been correct about the power of a balsam fir decoction to cure the very worst cases, this treatment would likely have risen to fame as a life-saving miracle, for diarrhea and dysentery routinely killed large numbers of children in the nineteenth century, especially during hot weather. Parents feared the euphemistically named "summer complaint," which was usually caused by *Shigella* or *Salmonella* bacteria or by intestinal viruses. But it was not going to be cured by a balsam fir decoction.

A major problem was that all early to mid-nineteenth-century practitioners and materia medica were ignorant of one stupendously important fact: microbes can make you sick. As a result, much of the practice of medicine and the use of plant-based drugs were based on ineffective treatments and ideas about disease that were dead wrong. To move forward, medicine needed a germ theory of disease—that is, the idea that specific microbes can invade the body and cause specific diseases. And to develop germ theory, researchers needed to be able to *see* the microbes that were causing skin infections, sore throats, pneumonia, tuberculosis, gonorrhea, dysentery, and other illnesses. By the early 1800s, versions of the microscope had existed for more than a hundred years, yet to that point no one had recognized the instrument's importance in explaining disease. This may have been due partly to the fact that the medically important group of microbes known as bacteria were exceedingly tiny. In order to see them clearly and thus begin to understand their significance, researchers would need much better microscopes.

The phenomenon of refraction explains why microscopes are able to create greatly magnified images of very small things. As a ray of light passes from the air into the glass of a lens, it is bent. The glass's refractive index value affects the degree to which the light is redirected; higher values indicate a more significant bend. For those working to improve microscopes in the late 1700s and early 1800s, refraction was both a friend and an enemy of sharp and clear enlarged images. It made microscopes (and telescopes) possible, but it also created problems that designers had to solve.

One major issue was that refraction could cause chromatic aberration, a problem related to lenses' ability to act as prisms. White light is a combination of a range of visible light wavelengths. When a beam of visible light passes through a glass lens, all of the wavelengths are refracted, but some are bent slightly more than others. This disperses the oncoming white light into a spectrum ranging from least-bent wavelengths, which are seen as red light, to most-bent wavelengths, which are seen as blue or violet light. You can see this effect illustrated on the cover of Pink Floyd's album *The Dark Side of the Moon*. But the prism creates challenges in microscope design because, on the viewer's side of the lens, the rays of different wavelengths meet at different focal points: blue meets at one, green at another, red at another, and so on. As a result, the viewer sees a chromatic aberration: a magnified image that is fuzzy and color-distorted, especially along the edges.[16]

Lenses with greater magnifying power also tend to have more severe chromatic aberrations. So as late eighteenth-century microscopists tried to build instruments with higher magnifications, they were bedeviled by the correlation between higher magnification and increased color errors. Some feared that the problem was insurmountable, but others continued to search for ways to reduce or eliminate the aberrations. By the early 1800s, instrument makers were combining lenses with different refractive indexes and complementary shapes, and this seemed to be leading them toward a solution. When lenses were combined into doublets or triplets, the separation of wavelengths created by the first lens in the combination was corrected as the dispersed wavelengths passed through the subsequent lens or lenses. In one widely adopted design, a plano-concave lens (flat on one side, inwardly curving on the other) made of flint glass with refractive index of about 1.62 was paired with a biconvex lens (outwardly curving on both sides) made of crown glass with a refractive index of about 1.52. Such combination lenses were called achromatic (without color) lenses because they mostly or completely eliminated the colored fringes along the edges of viewed objects.

Creating doublets and triplets had solved the color-error problem, but now there was another question: how should these lens combinations be held together? One possibility was to anchor them in the microscope's tube, leaving a thin layer of air between them, but this approach had drawbacks. As the microscopist Joseph Jackson Lister pointed out in 1830, a "dewiness or vegetation . . . form[ed] on the inner surfaces."[17] That is, the lenses could be clouded by condensation or mold. Moreover, as light rays left the glass of the first lens and entered the air in the gap, they would bend again. Depending on the angle at which the light had left the glass of the first lens, the rays might be scattered by reflections from the second lens. This would reduce image clarity and decrease the amount of light reaching the eyepiece at the top of the microscope. Lister worried that a gap between the lenses could result in the loss of "nearly half the . . . light from reflexion."[18] This could be critical at higher magnifications because high-powered lenses have smaller diameters than low-powered lenses do, and fewer light rays pass through them on their way to a viewer's eyes. The loss of light at high magnifications could make the image too dark to see, especially in the days before a specimen could be lit from below by bright electric lights.

Thus, it made sense to close the gaps by cementing the lenses together. But what could a microscopist use? The material would have to meet several criteria. First, it would have to be nearly or completely transparent; any glue that blocked the light passing through a lens would be worthless. Second, the adhesive should form a solid, permanent bond between pieces of glass. That eliminated slippery plant oils from contention: they were clear but would not provide enduring connections. Third, the substance's capacity to bend light as the rays passed through it should be equal or nearly equal to the refractive index of one of the two lenses it was gluing together: that is, the lens and the cement would act as a single transparent unit. That way there would be no additional bending of the light as it passed between the cement and the lens with a matching index. The lens and the cement would act as a single transparent unit. In contrast, if the glue's refractive index didn't match either lens, then there would be additional redirection of the light as it traveled between the lenses, as was the case when there was air in the gap.

And at this juncture, Canada balsam stepped onto the micros-copy stage. The resin checked every box. It was noted for clar-ity and transparency, possessed plenty of adhesive strength, and had a refractive index of about 1.53, an excellent match for the index of the crown glass used to make convex lenses. By the late 1790s, microscopists had begun using the resin as an optical cement in doublet and triplet lenses; and as the method spread, high-quality achromatic lenses became widely available to almost anyone interested in building microscopes. The juice from a tree blister had become an essential part of the solution to chromatic aberration.

The development of achromatic lenses was a leap forward in microscopy. But this success brought attention to yet another problem created by refraction: spherical aberration, which cre-ated a fuzzy image with smears of light resembling comet tails. Many of the achromatic lenses of the 1820s did a very good job of reducing chromatic aberration, but they were still produc-ing subpar images due to spherical aberration. This led some researchers to distrust the use of microscopes for biological research and led others to draw and report on structures that were artifacts of spherical distortions so did not actually exist.[19]

This problem fascinated Joseph Jackson Lister, whom I've already mentioned. Lister was an affluent wine merchant with time and money on his hands and a curiosity about microscopes. His interest in optics dated back to his childhood, when he had looked through air bubbles in windowpanes and discovered that the bubbles acted as lenses, improving his view of distant objects.[20] In the 1820s, as Lister studied the spherical aberration problem, he was not working in an intellectual vacuum. The Italian microscopist Gianbattista Amici had already made sig-nificant progress by combining achromatic doublet and triplet compound lenses in the same microscope and positioning them so that one lens would correct the spherical errors created by the other. Amici had experimented via trial and error, but Lister took a different approach. Like Amici, he used sets of compound lenses cemented together with Canada balsam and placed a certain distance apart in a microscope. But he was also looking for a theory that would explain why certain arrangements of these lenses could eliminate spherical aberrations. With such rules, he hoped to create formulas that would allow him to calculate how far apart achromatic compound lenses should be from one another to keep spherical errors to a minimum.

Lister tested his optical theories using lenses that he often ground himself, and his results confirmed that he could produce microscopes in which the spherical errors created by one achromatic lens were canceled out by another. He had also discovered principles that microscope makers could apply to reliably produce instruments that were far superior to those that had been made just a few years earlier. Lister secured his place in microscopy history with an 1830 journal article in the *Philosophical Transactions of the Royal Society of London*: "On Some Properties in Achromatic Object-glass Applicable to the Improvement of the Microscope." Writing in 1898, Dr. Henry Smith Williams gave Lister "chief credit for the directing those final steps that made the compound microscope a practical instrument instead of a scientific toy, . . . [c]ombining mathematical knowledge with mechanical ingenuity . . . [to produce images] highly magnified, yet relatively free from those distortions and fringes of color that had heretofore been so disastrous to the true interpretation of the magnified structures."[21]

Microscope makers in the 1830s used Canada balsam–cemented lenses, Lister's formulas, and the work of other innovators such as Amici to produce significantly improved instruments. These microscopes attracted the attention of a growing number of magnification enthusiasts, who used them to examine a wide range of biological objects. There were several ways to prepare specimens for study; but to enhance an object's appearance and preserve it for future examination, the microscopist would need to mount specimens on a slide using a medium that had several properties. The ideal agent would be transparent so that it wouldn't block the view of the specimen, sticky enough to anchor the specimen to a glass slide, able to adhere to the thinner glass cover placed on top of objects to be preserved, and capable of preventing air and water from reaching the specimen so that it wouldn't dehydrate or mold over time. The medium should also have a refractive index very close to that of the glass slides in the specimen sandwich. Ideally, the light would pass through the slide and specimen layers as if they were all made of the same solid transparent material. If it did not, one would have problems similar to those encountered when there was an air gap between lenses in doublets and triplets, and the image clarity would suffer.

Microscopists quickly identified the obvious material: Canada balsam. It was clear, sticky, and water-repellent, and its refractive index matched that of the crown glass used to make slides. In *A List of Two Thousand Microscopic Objects* (1835), the optician and naturalist Andrew Pritchard advised that many objects to be magnified were best studied by "immersing the object in Canada balsam or varnish, and pressing it between two slips of glass, so as to exclude air-bubbles."[22] In addition to the advantages I've already listed, Pritchard noted that balsam fir resin could render certain specimens more transparent, "allowing light to pass freely through [the specimen], exhibiting their structure, and presenting to the admiring spectator the most brilliant and superb colours." He continued to promote its use in his 1838 book, *Microscopic Illustrations of Living Objects*, co-written with Charles Goring. In this text, Pritchard described his studies of aquatic beetle larvae, stating that the insects' respiratory system "is seen to great advantage when mounted in Canada balsam." He enthusiastically recommended the resin as a "happy method" for "mounting and preserving insects."[23]

In the 1840s and 1850s, Canada balsam became increasingly popular as a mounting agent, and instructions for its use proliferated. John W. Griffith, a medical doctor and an amateur botanist, included directions for preserving slide specimens with Canada balsam in his 1843 paper "On the Different Modes of Preserving Microscopic Objects." He instructed readers to take a glass slide and "place upon it a small quantity of Canada balsam from [the] end of a piece of stick." They would need to give the balsam time to "perfectly, but slowly, melt and diffuse itself over the glass." Then they should lay the specimen in the center of the balsam puddle and, "if necessary, drop a small quantity more of the balsam upon it." Now they would need to set a "previously warmed glass slide upon the first; gently press them together; . . . [and] allow the balsam to solidify; the whole is then complete."[24] Voila! Budding scientists would now have a preserved specimen slide that could last for decades. Griffith did admit, however, that he might have accidently omitted some key steps: "One in the habitat of continually putting up specimens is apt to overlook mentioning certain minutiae, which, from use, he is hardly aware of performing, but which are essential to the perfection of this work."[25]

By the 1850s, the use of Canada balsam was so widespread that William Carpenter's *The Microscope and Its Revelations* (1856) included eight pages solely devoted to the subject of mounting objects in fir resin. Carpenter pointed out that these methods were suitable for a "very large proportion of those objects which are to be viewed by transmitted light."[26] His 1868 edition of the book expanded the Canada balsam section to nine pages and noted that there was now a market for instruments designed specifically for mounting specimens in fir resin. Carpenter highlighted one of his favorite's, James Smith's mounting instrument, a tool that included a platform for holding a glass slide and a knobbed arm that reached over the platform.[27]

To use the mounting instrument, the microscopist placed a specimen on a dry slide held by the platform and then carefully positioned a thin glass cover slide on top. The arm, acting like a clamp, gently held the slides and specimen in place. Then the user applied warm Canada balsam to one edge of the cover slide with a glass syringe, and the resin was drawn under the slide

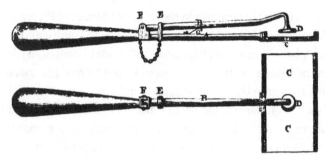

FIGURE 14. James Smith's mounting instrument. From William B. Carpenter, *The Microscope and Its Revelations*, rev. ed. (London: Churchill, 1868), 213.

and through the specimen by capillary action. An opening in the platform directly beneath the specimen allowed the user to warm the balsam on the slide with an alcohol lamp. This kept the melted resin moving across the slide, driving it into the specimen while driving out air bubbles. As Carpenter noted, the tool was "ingenious and convenient."[28]

As a lens cement and as a mounting medium for specimen slides, Canada balsam had become an essential part of microscopy. Now, in an 1854 issue of Charles Dickens's weekly magazine, Household Words, readers learned that fir resin alone had the power to create enlarged images. The revelation appeared in an anonymous piece titled "Catch-Pennies," a reference to cheap items designed to catch a buyer's eye. The author described "a man who sometimes stands in Leicester Street, who sells microscopes . . . made of a common pill-box." To turn the box into a magnifying instrument, the man had removed the bottom and replaced it with a piece of window glass to let light into the box, and a lens was held in "a small eye-hole . . . bored in the lid." The price was a penny per box at a time when a good-quality, high-powered microscopes sold for up to £25—the equivalent of 6,000 pennies.[29]

What could you see with a catchpenny scope? The author peered through the lens and "was surprised to find hundreds of creatures, [appearing to be] the size of earthworms, swimming about in all directions; yet [without magnification] nothing could be seen but a small fleck of flour and water." A second microscope provided a magnified view of a tiny animal that "show[ed] his impatience of imprisonment by kicking vigorously":

> Though I must confess to a shudder, I could not help admiring the beauties of construction in this little monster, which if at liberty, would have excited murderous feelings unfavorable to the prolongation of its existence. The sharp-pointed mouth, with which he works his diggings; his side-claws, wherewith to hold on while at work; and his little heart, pulsating slowly but forcibly, and sending a stream of blood down the large vessel in the centre of his white and transparent body, could also be seen and wondered at.

From this description, it appears that the author was admiring a louse.[30]

The writer "bought several of these microscopes determined to find out how all this could be done for a penny." He told readers that "an eminent microscopist examined the devices" and concluded that "the magnifying power was twenty diameters," similar to a powerful jeweler's loupe. At the time, a glass lens of similar power would have cost three or four shillings, the equivalent of thirty-six or forty-eight pennies. "How, then, could the whole apparatus be made for a single penny?" The answer: "The lens was made of Canada balsam, a transparent gum. The balsam had been heated, and carefully dropped into the eye-hole of the pill-box. It then assumed the proper size, shape, transparency, and polish, of a very well ground glass lens." In other words, the fir resin wasn't cementing lenses together. It was the lens, sold at a fraction of the price of a glass one. The "ingenious lens-maker" also kept costs down by operating a microscope assembly line staffed by himself, his wife, and three of his children. With free child labor, the man in Leicester Street could keep the price of his devices at a penny.[31]

Microscopes were becoming part of the culture of mid-nineteenth-century Europe. But what were microscopists studying with them? Almost everything: plants, animals, fungi, fossils, minerals—anything with features that might be enhanced by magnification. Some researchers turned their attention to human organs, examining healthy and diseased tissue, hoping to discover what had turned one into the other, though they made little progress in this regard. Others focused on a group of organisms called infusoria, whose name reflected their common presence in infusions of damp, decaying plants such as hay. It did not take long for biologists to realize that infusoria could be found in huge and diverse numbers in almost any body of fresh water. This grab bag of aquatic organisms included single-celled protozoans, photosynthetic algae, and tiny multi-celled animals such as rotifers. In *A History of Infusorial Animalcules*, Andrew Pritchard urged his readers to take the time to view a drop of pondwater under a microscope. They would be rewarded by the sight of "beautiful and varied forms," "admirable in structure," to be observed "with as much liberty and pleasure as the mightiest monster of the deep."[32]

Improved by Canada balsam, the microscopes of the 1840s and 1850s expanded the study of single-celled protozoans such as amoebas. However, there was one group of single-celled organisms that researchers continued to almost completely overlook: bacteria. Why weren't these microbes getting any love from mid-nineteenth-century microscopists? The answer is unclear, but there are several possibilities. As I've noted, bacterial cells are tiny, much smaller than the microscopic protozoans. Despite improvements in the scopes, they were not easy to see. Nor were they much to look at once you did see them. Most bacteria are either spherical or rod-shaped and, under the microscopes of the day, would have seemed plain and almost featureless. Compared to the "beautiful and varied forms" of the infusoria, bacteria were rather dull. It would have been difficult to get rave reviews for journal articles featuring drawings of uniform spheres and rods. It would also have been easy to conclude that there were only a few types or species in this group of microbes. As far as anyone knew, they were of little importance.

Today we automatically assume that certain types of bacteria can make us sick or even kill us. Improvements in microscope technology might have lead researchers to this knowledge as early as the 1830s. But for them, the idea that something as tiny as a single-celled bacterium could kill a human being was almost unimaginable. Bacteria were too small, too simple, too insignificant to be responsible for skin infections, blood poisoning, scarlet fever, diphtheria, pneumonia, tuberculosis, dysentery, and gonorrhea. True, if researchers examined the pus or mucus produced by these diseases under a microscope, they could see bacteria. But they could also see that strong and vigorous people were routinely loaded with what looked to be exactly the same bacteria, so there was no correlation between the presence of bacteria and illness. Besides, the medical establishment had already embraced certain ideas about disease. Most practitioners accepted the idea that illnesses ensued from the transmission of some vaguely defined nonliving agent—maybe a poison or a chemical catalyst—via corrupted air (known as miasma) or by contact with another person.

Still, a few researchers raised the possibility that bacteria could cause illness. Of these, the best known was not a medical doctor but a chemist named Louis Pasteur. Pasteur didn't begin his research career with the goal of becoming one of the first people to promote the idea that microbes cause disease. He followed a winding research path to this conclusion, relying on his microscope, which was always nearby. Early in his research life, he had taken a strong interest in the chemical changes produced by fermentation, and his investigations in the 1850s led him to conclude that it was the result of microbial activity. Today this seems obvious, but at the time it was a new idea. In the course of his studies, he noticed that some types of microorganisms would ferment sugar into alcohol while others would turn it into lactic acids. That is, he discovered that different microbes produced different chemical outcomes. Later, when asked to solve the problem of sick caterpillars in French silkworm colonies, Pasteur found that one type of microbe was causing a disease called pébrine while another was causing a disease called flacherie. Now that Pasteur was learning that different types of single-celled organisms could cause different types of changes in sugary juices and insects, he grasped that there was a broader principle at work here. Perhaps different types of microbes, including different types of bacteria, could also cause different types of diseases in humans.[33]

In proposing that bacteria and other microbes cause human disease, Pasteur was describing germ theory. He was not the first or only person at the time to connect germs and illness, but he was better than most at promoting and testing the idea. His research was limited by a lack of bacteriological methods, which would not be developed until the 1870s and 1880s, and by the fact that he was not a medical doctor and thus could not treat or work with human patients and human diseases. Still, he could disseminate his ideas and hope that a medical practitioner might accept and act on them. And in 1865 that's exactly what happened. A young surgeon in Glasgow, Scotland, embraced Pasteur's ideas, put them into practice, revolutionized surgery, and provided compelling evidence that germ theory was valid.

In the mid-nineteenth century, the field of surgery was in a peculiar situation. For thousands of years, surgeons and patients had dreamed of finding a way to perform operations painlessly. Those dreams turned into reality in the late 1840s, when researchers discovered that ether and chloroform could function as general anesthetics. Now surgeons could operate freely, and patients would feel no pain during the procedure. This was wonderful news. Unfortunately, however, surgeons had no idea that microbes caused disease. And now, with the patient unconscious throughout the process, surgeons could operate for longer periods of time and attempt operations they had avoided before, thus allowing environmental bacteria to enter the bodies and cause infections. The surgeons themselves contributed to the problem by working with unwashed hands and filthy instruments. Why shouldn't they wear their dirtiest clothes to operate? Everyone knew that bacteria didn't cause infection, right?

The results of these practices were dire. Surgery was routinely followed by life-threatening infection, and postsurgical mortalities of 50 percent or higher were the norm. We think of "the operation was a success, but the patient died" as a cliché, but it was a reality in the nineteenth century.[34] This fact greatly troubled the young surgeon in Glasgow. He wanted to end these high mortality rates; and when he encountered Pasteur's ideas, he saw a solution: "We find that a flood of light has been thrown upon this most important subject by . . . M. Pasteur, who has demonstrated by thoroughly convincing evidence that it is . . . the germs of various low forms of life, long since revealed by the microscope, . . . to be [the] essential cause [of infection]."[35] To the surgeon, it was clear that disease, including postsurgical infection, was caused by "germs of various low forms of life." And not only could germ theory explain surgical infections, but it also pointed to a way to stop them. The surgeon realized that if he could kill germs in the air and on hands, instruments, and other surfaces contaminated with microbes, then he could prevent disease. He could save lives. To do the killing, he turned to a chemical called carbolic acid, better known today as phenol. While carbolic acid did cause some damage to human skin, he believed that the damage was outweighed by the chemical's strong antimicrobial activity. With carbolic acid in hand, he had what he needed to destroy the germs before they could destroy a patient. (Today, surgeons have replaced phenol with less harsh preoperative skin antiseptics.)

The surgeon decided that he would first try out his carbolic acid theory on patients with compound factures. While these cases were not planned surgeries, they were similar in that the skin was torn open (by bone instead of a surgeon) so microbes in the air or on the skin were able to enter the body and cause blood or deep tissue infections. Previously, amputation had been the only way to prevent a fatal case of bacterial gangrene, and even then patients often died. The surgeon in Glasgow was eager to test Pasteur's new germ theory. If it prevented infections, it would save limbs and lives and offer strong evidence that these methods could improve a range of surgical procedures.

Now all the surgeon needed was an experimental subject. He did not have to wait long, for the streets of Glasgow, packed with carriages and wagons, were a dangerous place for pedestrians. On August 12, 1865, eleven-year-old James Greenlees stepped into the path of a fast-moving cart. His left leg was crushed under one of the wheels, resulting in a compound fracture of the tibia. Greenlees was taken to the Royal Infirmary, where the surgeon had to make a decision: amputate or try carbolic acid? In an economy dependent on manual labor, the loss of a leg could mean poverty for life. But if gangrene were to set in, the boy would likely have no future at all. The surgeon decided to trust in germ theory and try to save the leg. He gave the boy chloroform and then cleaned and washed the wound with carbolic acid to kill the bacteria that had already entered the gash. He set the broken bone with "lateral pasteboard splints padded with cotton wool" and covered the wound with "a piece of lint dipped in liquid carbolic acid."[36] Over subsequent days and weeks, the surgeon continued to apply carbolic acid to the slowly healing wound. Six weeks and two days after the accident, James Greenlees left the hospital with both his leg and his life. Killing the bacteria had worked.

What made this particular surgeon so receptive to Pasteur's ideas when so many other practitioners were ignoring or deriding the ideas of a mere chemist? At this time, most doctors and surgeons had never looked through a microscope or considered the possibility that the tiny cells on the specimen slides might be killing their patients. But this surgeon had an unusual history. As a boy, he had been surrounded by up-to-date microscopes and by people who were eagerly using those Canada balsam–cemented lenses to view specimens mounted in Canada balsam puddles. He may even have prepared his own slides, using Canada balsam to fix and preserve them. To this surgeon, the microscopic world was not a foreign land, so it was easy for him to believe that there was an invisible country alive with organisms.

The surgeon's name was Joseph Lister, and his father was Joseph Jackson Lister, the famous microscope theorist, author of numerous microscopy papers, and a founding member of the Microscopical Society of London. In his household, J. J. Lister had created the environment that led his son to a life-long love of microscopy, and the elder Lister's efforts to eliminate spherical aberration had helped to create the high-quality instruments that had inspired the curiosity of a child. As an anonymous contributor to the British Medical Journal wrote in 1898, "it is hardly too much to say that the discoveries which have made the name of the son immortal could not have been made without the previous discoveries of the father."[37]

In Joseph Lister's 1867 article "On a New Method of Treating Compound Fractures, Abscesses, Etc.," James Greenlees is presented as "CASE 1."[38] After this success, Lister successfully treated several more compound fractures and then moved on to using carbolic acid to surgically treat abscesses. Over time, he expanded his germ theory–based methods to other types of surgery, even performing a mastectomy on his sister after the appearance of a cancerous tumor. In all cases, deaths from postsurgical infections declined dramatically. Lister published frequently and traveled in Europe and the United States, sharing his transformative approach to operations. It took most of two decades for his methods to become widely accepted because many still rejected germ theory and accepting what Lister was teaching would mean admitting that surgeons had been unintentionally responsible for countless deaths from infections. But the benefits and effectiveness of Listerism were undeniable. Today's aseptic surgery—its disinfected operating room; sterilized gloves, gowns, and masks; and pre-op body surfaces scrubbed with antiseptic—descends directly from Lister's carbolic acid and young James Greenlees's leg.

Joseph Lister changed surgery, but he also offered powerful evidence that the germ theory of disease was correct. As an increasing number of doctors and scientists accepted the theory, they fixed their sights on those neglected and overlooked microbes called bacteria. In Germany, the physician and microbiologist Robert Koch and his colleagues developed new methods for staining bacteria, for creating pure cultures with a single species of bacteria, and for proving that a specific species of bacteria had caused a specific disease. By the late 1800s, researchers were proving that anthrax, tuberculosis, cholera, typhoid, pneumonia, strep throat, skin infections, and blood infections were caused by bacteria. Germ theory also explains why Alexander Mackenzie's salve of Canadian balsam, wax, and tallow may have healed his patient's hand: the terpenes in the resin likely killed the bacteria that caused the infection.

Balsam fir resin played multiple important roles in one of the greatest discoveries in nineteenth-century medicine, and its use in microscopy continued to support new research. Late in the century, when Koch and other microbiologists were linking bacterial species to different human diseases, they used Canada balsam to preserve slides of infected human tissue containing stained, disease-causing bacteria. The researchers' surviving storage cabinets are filled with specimens sandwiched between thin, yellowing, and often cracked films of fir resin. If Canada balsam was not the lead character in the story of germ theory, it certainly earned a nomination for best supporting performer.

Chapter 7

Pitch Pine Resins in Medicine, Naval Stores, and Colonial Conflicts

IN THE FEBRUARY 18, 1774, issue of the *New Hampshire Gazette*, the following notice was posted by the "Committee for Tarring and Feathering":

> Many People being Apprehensive that there will be a Difficulty in preventing some individual Persons from selling TEA, even tho' this Town should Vote against it; We think proper to declare to YOU [who wish to prevent the sale of tea] that you need not doubt of your Resolutions being carried into Execution; for we whose Names are hereunto subscribed, will engaged that no Persons in this Town, Great or Small, Rich or Poor, shall dare to counteract your laudable Designs.
>
> The Committee for Tarring and Feathering:
> Thomas Tarbucket, Peter Pitch, Abraham Wildfowl,
> David Plaster, Benjamin Brush,
> Oliver Scarecrow, Henry Hand-Cart[1]

What had prompted "Thomas Tarbucket" and friends to post this announcement? In 1773, the British government had passed the Tea Act, which levied duties on tea imported into the American colonies. Angry colonists called for a boycott on British tea, but many merchants were reluctant to take part as a boycott would cut into their sales. The Committee for Tarring and Feathering was not sympathetic to this point of view. In this public notice, it was letting local sellers know that if they failed to comply with the boycott, the committee would punish them: first, by coating them with the resinous product of a local conifer species; then, by rolling them in feathers, no doubt supplied by "Abraham Wildfowl."

The conifer was the pitch pine (*Pinus rigida*), a species that has played many roles in the history of New England: as a source of light, medicine, and marine products; as a trigger in a border dispute between colonies; and as an adhesive in the political ritual called tarring and feathering. All pine species of the Acadian Forest ecoregion produce defensive resins, but pitch pine is humans' most significant source in this region. These gummy pine terpenes make the wood rot-resistant and easy to ignite, and they are the source of turpentine, tar, pitch, and rosin, a group of products known collectively as naval stores. Pitch pines do not grow in most of the Acadian Forest ecoregion, only in the southern part, especially in sandy, dry, acidic, less fertile soils and on rocky ridges where other tree species struggle to compete. The pines often colonize open sites such as abandoned farm fields, land cleared by logging or fire, and sandy plains of glacial origins. In locations where poor soils and frequent fires select against other tree species, pitch pines can dominate a forest for decades and may live for up to two hundred years while reaching a height of ninety feet.

Despite pitch pine's flammable resins, fire tolerance is its superpower. The species is one of the most fire-tolerant in New England forests, and it benefits from periodic blazes. The trees are protected by fire-resistant bark, which on their trunks may be up to two inches thick a foot above the ground. If a fire destroys the top of the tree, the pine has the capacity to sprout from buds at the base of its trunk and on its root crowns, a rare trait among conifers in the Acadian Forest. Young trees often grow with a slight crook at the base of the trunk, and the dormant buds on the underside of that crook are buried in soil that insulates them from fire. Pitch pines also produce epicormic buds along the trunks; these are bristly, needle-bearing bumps that can sprout to form new branches after a fire. In addition, the trees have serotinous cones—that is, cones that remain closed for an indefinite period of time after they mature. The seeds inside are thus protected during fires, and afterward they are released into a world with less competition, plenty of sunlight, and a stock of nutrients in the ashes that would have otherwise been in short supply.[2]

Given its range and adaptations, pitch pine was available to New Englanders even in the earliest days of colonization. The tree's resins burn well and keep a steady flame, so it was a useful source of light. Colonists called it candlewood or torch pine and would pack slivers from the resinous heartwood (known as fat wood) into lamps and ignite them to create "splint lights." They would also burn the tree's resin-rich knots. John Winthrop Jr., the governor of the Connecticut Colony and a son of one of the founders of the Massachusetts Bay Colony, wrote that the colonists would split the knots "into small shivers, about the thickness of a finger, or thinner; and those they burn instead of candles, giving a very good light, and they call it candlewood." But burning a knot did have its drawbacks: "Because it is something offensive, by the much sullying smoke that comes from it, they usually burn it in the chimney-corner upon a flat stone or iron, except occasionally a single stick in their hand, as there is need of light to go about the house."[3]

The antimicrobial properties of pitch pine resin made the wood resistant to decay, so its boards were valuable for house sills that were in contact with damp soil, for waterwheels in mills, and for underground pipes that supplied towns with water. Even very early colonists took advantage of these features. For instance, archeologists found proof of pitch pine's extreme durability and rot-resistance at the site of an early settlement known as the Popham Colony. Established in late 1607 near the mouth of the Kennebec River, it was the first English colony in New England in which the inhabitants spent the winter. However, their leader,

George Popham, died in February 1608, and his successor, Raleigh Gilbert, decided to return to England to cash in on a new inheritance. The other colonists also decided to leave. They had failed to establish good relations with the local tribes, a prerequisite for any profitable fur-trading operation, and no doubt were not looking forward to another harsh Maine winter.

Popham Colony had been an ambitious undertaking, and excavations in the 1990s and early 2000s revealed that pitch pine had been an important building material. It was used for the supporting posts of major buildings inside the colony's Fort Saint George, including the seventy- by twenty-foot storehouse, the largest structure in the settlement. In several storehouse postholes, archeologists found the decay-resistant remnants of four-hundred-year-old wooden posts containing resin-filled knots, a sign that the wood was pitch pine. The post remains were up to two feet long, flat on the bottom, and most were cut to be roughly square or rectangular in cross-section. One preserved post bottom was thirteen inches across on one side and likely came from a pitch pine that was at least eighteen inches in diameter, about a hundred years old, and as tall as sixty or seventy feet. The colonists had also used pine for the wall sills running between the posts. Again, the presence of resinous pine knots showed that the species was pitch pine. These remnants are unusual in excavated North American settlements from the 1600s. Archeologists more often find dark-stained soil where posts once stood but no intact wood in the holes.[4]

Although terpene-saturated pine is handy for torches and rot-resistant projects, most other uses require the terpenes to be separated from the wood. As mentioned, a number of these commodities are known collectively as naval stores because they were heavily used in marine applications. These stores are divided into two groups: (1) turpentine and rosin made from resin extracted from a live tree after the bark has been cut and (2) tar and pitch produced from resin driven out of the wood by heating the knots, branches, and trunks of dead pines. During the colonial period,

New England's pitch pines were an important source of these stores, but eventually almost all American-produced turpentine and tar would be derived from two southern pine species: yellow or longleaf pine (*Pinus palustris*) and loblolly pine (*Pinus taeda*).

Turpentine production began with the extraction of oleoresin from a live pine tree. In a process known as boxing, a small box-like hollow was cut out at the tree's base, and the bark was slashed above the box to create a set of parallel, downward-pointing chevrons. Resin oozed out of the damaged bark, flowed into the hollow, and was eventually scooped into barrels for transport. This gum was sometimes referred to as crude, or raw, turpentine, which can be a little confusing because today the word *turpentine* usually refers more narrowly to an oily liquid produced by distilling pine tree resin. But in the past, the word could refer to any undistilled or unprocessed resin or gum that oozed out of any cut conifer species. Crude turpentine could be used as an adhesive, a sealant, a wood coating, or a base for inks. However, most of the collected resin was destined to be heated in a distillation apparatus to divide the resin into oil of turpentine and rosin.[5]

In distillation, pine sap is heated, and its components are separated on the basis of the temperatures at which a given terpene boils or easily vaporizes. At certain temperatures, this produces a vapor dominated by monoterpenes with some sesquiterpenes mixed in. These fumes are condensed on cool surfaces, forming a collected liquid called oil of turpentine or spirits of turpentine— what we today simply call turpentine. Because distillation uses heat and creates highly flammable vapors and liquids, fires were common in early turpentine stills.

Turpentine was considered a naval store because seamen on wooden ships used it to waterproof the hemp rope rigging and added it to the paint and varnishes they used to waterproof the decks. Today, it is still sold as a solvent for paints and varnishes and as a source of aromatic monoterpenes for fragrances. But at one time it also had multiple uses in medicine. Many nineteenth-century doctors, supported by their materia medica, promoted the idea that crude resin turpentine and distilled oil of turpentine had healing properties. Although both commodities were seen as valuable, most medical treatments called for the use of the latter, often referring to it as *oleum terebinthina*, perhaps to make it sound more mysterious and potent. Oil of turpentine or lotions containing it were applied externally to joints to treat pain and swelling; to the neck and chest to ease sore throats, bronchitis, and asthma; and on skin infections such as *Staphylococcus*-filled carbuncles.[6] It's possible that turpentine's antimicrobial terpenes, like balsam fir's, had some ability to treat superficial infections. But *oleum terebinthina* also caused a burning sensation and could significantly irritate and blister skin. Oddly, this skin damage was sometimes the goal of the treatment, as doctors thought it would act as a counter-irritant." According to disease theories of the time, a counter-irritant would offset or draw away an internal "irritant" responsible for a disease; and by balancing or by shifting the location of the disease-causing matter to the surface, turpentine would cure the disease. In Gunn's *New Domestic Physician*, John Gunn recommended treating cases of rabies by applying oil of turpentine to the feet and legs. He held that this would "make use of powerful counter-irritation along the whole course of the spine, by the application of strong stimulating liniments." Gunn accurately described rabies as "terrible and always to be dreaded," but he must have known that nothing was going to prevent death once its symptoms appeared.[7]

Even before they understood the role of microscopic bacteria, doctors were aware that wounds could be colonized by very visible maggots (fly larvae), and some also applied turpentine to infested wounds to treat this problem. In his *Personal Recollections of the War of 1861*, Charles Fuller, a lieutenant in the Sixty-First New York Infantry Regiment, recorded an account of turpentine therapy for maggots. Fuller had been wounded in the thigh during the second day's fighting at Gettysburg. About a week after his leg had been amputated, he experienced a "crawly feeling" at the end of the stump. He brought this issue to the attention of a passing doctor, who undid the bandages and found "a large number of full-grown maggots in the wound." Fuller was horrified, but the doctor assured him "that the wigglers didn't amount to much in that place." And the doctor had a solution: "He diluted some turpentine, took a quantity of it in his mouth and squirted it into the wound, and over the stump. It did the business for the intruders, and I had no more trouble of that sort."[8]

Applying turpentine externally could seriously damage skin tissue, and some of the terpene molecules could pass through the skin and enter the body, but the treatment was unlikely to kill anyone. Internal use, however, was much more dangerous. Pine terpenes kill pathogens by damaging their cell membrane; and as I noted in chapter 7, the cells in our body also have membranes vulnerable to terpene damage, especially internally. Even at low doses, consuming oil of turpentine could cause nausea, vomiting, and severe diarrhea—though, in some cases, doctors prescribed turpentine for exactly these results. Until the advent of germ theory in the 1860s and 1870s, many believed that diseases were caused by imbalances that could be corrected with cathartics such as *oleum terebinthina* that would induce vomiting and diarrhea. Nineteenth-century doctors knew that drinking or inhaling large doses of oil of turpentine could lead to mental confusion, convulsions, coma, or death. But even though this fact appeared in medical texts, practitioners still recommended its internal use for an extensive list of ailments.[9]

Because they recognized that the oily liquid increased urine flow, doctors often prescribed turpentine for excretory-system problems ranging from kidney ailments to urinary-tract infections, including gonorrhea. It was often used for digestive-tract diseases such as dysentery, typhoid fever, and cholera, although increased diarrhea was the last thing these patients needed, as this could hasten death by dehydration.[10] Doctors varied in their use of oil of turpentine for the above conditions, but everyone agreed about one thing. It was an especially effective way to rid the body of tapeworms. As Robert Christison wrote in his *Dispensatory*, the oil's "most unequivocal therapeutic action is that of a vermifuge, [for] worms generally, but above all in tape-worm. In this variety [of worm], which resists most other anthelminthics, oil of turpentine taken inwardly is the most certain of all remedies."[11]

Oil of turpentine was even prescribed for minor problems such as flatulence. Never mind the poisoning, we must stop the passing of gas.[12] But fortunately its very unpleasant taste and the burning sensation it causes in the esophagus and stomach may have limited its internal use. In an 1840s medical text, John Pereira endorsed repeated doses of turpentine as a treatment for childbed fever, a bacterial infection that could follow the birth of a child and kill both mother and infant. Still, he also noted that the "unconquerable aversion which patients have manifested to [turpentine] had precluded its repetition." Apparently, some patients thought drinking oil of turpentine was worse than the disease itself. Pereira did have a solution to this problem: turpentine enemas. It seems he was determined to force *oleum terebinthina* onto his patients, and the fact that his treatment for childbed fever was both worthless and dangerous does not appear to have been a deterrent.[13]

Distillation of pine sap also generates a product called rosin. Before heating, monoterpenes in pine sap act as solvents that dissolve and suspend diterpenes and diterpenoid acids and keep the resin in a viscous liquid form. But as distillation removes monoterpenes to make oil of turpentine, the diterpene-rich material left in the bottom of the still becomes more solid. This is rosin, and seamen used it as a waterproofing agent. It's also found in printing inks, varnishes, soap, and sealing wax. Violinists and other string players apply it to the horsehair on bows to improve its friction on the strings and create clean vibrations, and powdered rosin in bags helps pitchers get a better grip on baseballs.

As mentioned, turpentine and rosin are made by heating

terpene-rich pine resin after it has been passively drained from
live trees. But other pine products are produced by heating pine
in kilns to cook the resin and drive it out of the wood. Tar is
the initial product, and pitch is made by cooking the tar further.
Traditionally, there were several approaches to securing material
for the kilns. Tar makers could gather wood from the remains
of dead pines. Or they could cut down live pine trees with the
bark still intact. Or they could remove much of the bark from
live trees and then wait for one to three years for the injured
tree to fill the trunk with defensive resin and then harvest the
wood. The final tar product could vary significantly, depending
in part on which gathering method was used.

When following the first method, tar makers gathered pine
remains that contained a high proportion of the heartwood,
formed by the tree's oldest inner rings. The heartwood may contain two to three times as much resin per pound as the younger
outer sapwood does. In addition, these high resin concentrations
make the heartwood rot-resistant; so when branches break off
or a tree dies and falls, the terpenes preserve the heartwood
for many years while the bark and sapwood slowly decompose
and disappear. An old-growth forest may be filled with an accumulation of heartwood in the form of the durable interiors of
branches, trunks, stumps, and the pine knots that contain the
heartwood, which forms at points where tree limbs branch off
from the trunk.[14] Such woodlands were common in New England
when the colonists arrived in the 1600s.

The second approach to tar making entailed felling a live tree
and cutting it into one- or two-foot-long sections for the tar kilns.
But there were notable problems with this approach. Fresh logs
were primarily composed of lower-resin sapwood, and a live tree
left standing could be a source of potentially more profitable
turpentine or lumber.[15] Thus, using live trees for tar might not
make sense unless the harvester could significantly increase the
resin concentration in the heartwood.

Thus, tar makers developed a third approach, a centuries-old technique known as the Swedish or Finnish method, named after the Swedish merchants who had popularized it and after the early source of the tar, Ostrobothnia on the west coast of Finland, which was controlled by the Swedish Empire for much of the seventeenth and eighteenth centuries.[16] In Europe, tar makers began the process by barking a live Scots pine (*Pinus sylvestris*), the same species whose trunks were exported for ship masts. They peeled the bark from around most of the trunk, starting at the ground and working up to a height of six or eight feet, but left a vertical strip of bark about four inches wide on the north side of the tree to maintain the connection between the roots and the rest of the tree. This would keep it alive for a few more years. The injured tree responded by producing more protective resin and transporting both new and stored resin to the debarked trunk. Terpene-rich pine gum oozed from resin ducts located in the outer sapwood and along the edges of the cut bark, covering the surface of the trunk. After a period of time, barkers could remove additional horizontal bands of bark above the original cut to induce more resin production in the newly exposed sections. Between one and three years after barking, the tar makers would have wood containing far more resin than the trunk of an undamaged tree. Because the resin had been produced from a living tree, the final product was known as green tar. In some ways, this method paralleled turpentining in that the trees were damaged to bring both stored and newly produced resin to the surface. However, for tar making, the trunk was cut into short sections, and the resin was separated from the wood in a kiln. The method generated high-quality tar, and buyers preferred the product. But it also killed a huge number of live trees and ultimately destroyed entire forests of Scots pines.[17]

Regardless of the wood's source, extracting tar required a kiln. Most were circular or oval-shaped, ten to twenty feet in diameter, and sunk partially into the earth or dug out of the slope of a hill. The Finnish word for tar kiln, *tervahauta*, literally means "tar tomb." The base might be lined with clay or stone and was cone-shaped to funnel the exuded tar downward into a drain at the point. Tar makers packed the kiln with resin-rich knots, branches, or logs; if logs were used, they were arranged like spokes, with their ends pointing toward the funnel. Once filled, the kiln was capped with earth, sod, or clay and then ignited through a hole at the top. Heat was required to extract the resin, so some of the wood had to burn to warm the kiln's contents and some oxygen had to circulate to sustain the fire. However, with a strong airflow, the burning would accelerate, and too much of the wood would overheat, ignite, and burn vigorously instead of smoldering. The valuable tar would go up in smoke, leaving tar makers with a pile of nearly worthless ash. So workers carefully controlled the airflow via the opening at the top and by holes along the sides of the kiln. The goal was a Goldilocks-like balance: just the right amount of heat and oxygen. When operating properly, the kiln cooked and charred the resin, darkening its color. The heat also eliminated some of the smaller, easily vaporized monoterpenes, increasing the concentration of the more viscous, less volatile diterpenes and diterpenoid acids and thickening the resinous extract. As the tar warmed, it would ooze out of the knots and logs, down into the kiln's bottom, and then out of the kiln into troughs to be captured in containers for transport and sale.[18]

Pine tar was used as a wood sealant and preservative, as an antiseptic to treat skin ailments, and as axle grease for wagons; many wagons carried buckets of pine tar so that it could be applied as needed. But it was particularly valuable on naval and commercial vessels. Sailing ships require enormous amount of rope, much of which was made from strands of hemp that had been "tarred in the yarn"—that is, treated with hot tar. It was a delicate process. The tar had to be hot enough to flow into the strands; but if it was too hot, the fibers would burn and lose strength. Once the strands were tarred, they were woven into ropes that eventually became rigging, hawsers (mooring lines), and anchor cables. The tar waterproofed the ropes and prevented saltwater corrosion; without it, they would rot quickly due to constant humidity and sea spray. Tar could also be used to stiffen ropes so that certain types of rigging could serve as ladders. But the tar infusion did not last forever, so sailors kept barrels of tar on board and periodically recoated the rigging, a chore that forced them to swing dangerously in the air clutching brushes and buckets. Tar was also a component in the stringy, hemp-based caulking material called oakum, which shipbuilders and sailors drove into the seams between planks as a sealant and preservative. Sailors even coated their canvas hats and clothing in tar to create a waterproof material known as tarpaulin.[19]

Tar was so strongly associated with mariners that, by the 1700s, English sailors were referred to as "jack tars" and "jolly tars"; the word *tar* has since become a general synonym for *sailor*. In the 1700s and early 1800s, naval press gangs often roamed the streets of British port towns, looking for men to kidnap and force onto Royal Navy vessels. These armed gangs preferred to grab men with prior experience, and they looked for telltale tar stains to identify those who had worked at sea. One such victim, a sailor named Robert Hay, recalled that he was walking the streets of London in 1811 when he was tapped on the shoulder by a man in seaman's dress. The stranger asked, "What ship?" Hay knew the man was from a press gang, so he declared that he "was not connected with shipping." Nonetheless, he was dragged away by "six or eight ruffians" and brought "into the presence of a lieutenant who questioned me as to my profession, and whether I had ever been to sea." Hay answered evasively, but his fate was sealed when his hands were found to be "discoloured with tar." Though he was pressed into the Royal Navy, he was able to escape and swim to shore before the ship could leave English waters.[20]

Pine tar was also the starting substance for pitch (as in pitch pine), which was made by boiling tar in large iron kettle. This evaporated almost all of the remaining monoterpenes that had acted as solvents and kept the tar in a slightly liquid form. The resulting pitch was thicker and almost solid at room temperature.

It was also very dark in color, and this characteristic might be the origin of the phase "pitch black." When pitch was reheated and spread over a surface, it cooled and hardened to form a nearly solid watertight seal, and it was often applied to ships' hulls to plug leaks. It also repelled wood-drilling *Teredo* mollusks, known as shipworms, that could tunnel into wooden hulls and sink a ship. Sailors stored barrels of pitch on board so that it could be reapplied as needed. To perform this task, captains careened, or beached, their ships at low tide, leaving the hulls exposed to the air. After heating the pitch until it was runny, sailors would smear the goop on the hulls. Then the ship would refloat on the incoming tide. Sometimes, however, a hull seam would begin to leak badly before a ship could be careened, meaning that the pitch would have to be reapplied at sea. A deckhand would be lowered over the side with a supply of heated pitch and instructions to brush the material into the leaky seams. If the leak were lower down on the hull, sailors on deck might have to set the sails and shift heavy objects to tip the ship enough to expose the section in need of repair.[21]

Pitch pines and naval stores played significant roles in New England's history, especially during the colonial period. As with the white pines, the British Empire saw the region's pitch pines in terms of their potential utility to the Royal Navy. Its policies promoting the production of turpentine pushed the colonies of Connecticut and Massachusetts into a conflict over colonial borders and control of pitch pine stands in the Connecticut River valley. Likewise, pine tar became a political tool when mobs used tarring and feathering to intimidate British officials and others who disagreed with them. The old-growth white pines may have been more noble and impressive, but the scraggly *Pinus rigida* would also leave its mark on New England culture.

The pitch pine's potential was recognized early in the seventeenth century. In the 1630s, Thomas Morton, an early Massachusetts colonist, noted in his book *New English Canaan* that "there is infinite store [of pines] in some parts of the

Country. . . . I have travelled 10. miles together, where [there] is little, or no other wood growing. And of these may be made rosin, pitch, and tarre." He worried that if the English could not import these commodities from other countries, "our Navigation would decline." But if relations with those sources should sour, "then how great the commodity of it will be to our Nation, to have it of our owne [from New England]."[22]

In the early 1660s, Governor John Winthrop Jr., sent an extensive report to the science-focused Royal Society of London explaining where and how New Englanders were using pitch pines to make tar. He noticed that the "greater quantity" of the pines were to be found in poorer soils and on rocky hills, an observation that is consistent with what we know about this species's ecology. He also identified the lower Connecticut River valley as the main region for tar production, primarily in locations "above fifty miles up the river, where there be great plains of those pines on both sides [of] the river, [and] up into the land from the river side."[23]

Winthrop observed that New Englanders of the 1660s were using pine knots "impregnated with that terebinthine or resinous matter" as a major source of heartwood for tar production. Many of the knots came from "trees which have been blown down, . . . [and] been lodged there so long time . . . that the whole body of the tree, and all the boughs and roots thereof, are rotted, and only the knots of those boughs left." He explained why the colonists chose to make tar from pine knots scattered on the forest floor instead of recently harvested trees:

> It must not be conceived, . . . that only those knots, separated from the body of the tree by devouring time [which rotted the rest of the tree], are the only material out of which tar can be made; for there are in that country millions of living trees, that have the same sort of knots full of such turpentine, . . . but the labour of felling the trees, and cutting out those knots, would far exceed the value of the tar, . . . especially in that country, where labour is very dear.

In contrast, it didn't take much time or effort to gather pine knots in old-growth forests, which is why mid-seventeenth-century New England tar kilns cooked pine knots.[24]

As Winthrop noted, much of colonial New England's tar production was centered in the lower Connecticut River valley in the area around the township of Windsor in Connecticut Colony. He described the plains as "barren," which likely meant that the sandy soils were not good for farming. He may have been correct about the agricultural potential of this land in the 1600s, but the soils were perfect for the pines because they were able to outcompete other tree species in habitats low in plant nutrients. Valley inhabitants had already been using these pines to produce modest amounts of tar and pitch; but as the century drew to a close, they began to take an increasing interest in the possibility of creating a different naval store: raw turpentine. (They would leave the subsequent distillation which yielded oil or spirits of turpentine to workers in Boston or London.)

Although pitch pines were abundant, turpentining was not necessarily the clearest path to economic success in the lower Connecticut River valley. The region was thinly populated, meaning that labor was expensive and risked consuming whatever profits might be made from collecting pine resin. Land, however, was readily available and relatively cheap, so most settlers could earn more money from farming than from tapping pine trees. Colonists mostly thought of turpentining as a part-time source of supplemental income during slowdowns in the farming cycle. The also relied on pitch pines for lumber; and if a sawmill were nearby, they were likely to harvest the pines for building materials rather than leave them standing for resin collection. Turpentining created a sticky residue on the bottom of the trunk, which gummed up sawmill blades, and drained the resin from the top of the tree, leaving it brittle and prone to decay. As a result, most resin gathering was done in dense pine stands located too far from the mills to make harvesting for lumber profitable but close enough to roads and rivers to make it worthwhile to wrestle barrels of resin to waiting ships.[25]

Still, the turpentine business had advantages. Collectors did not need to invest time and money in building tar kilns or sawmills. And unlike tar making and lumbering, turpentining did not require colonists to spend long periods alone in the woods—a factor of great importance at a time of near-constant conflict among the English, the French, and various Algonquin and Iroquois tribes. In a 1704 raid on the English settlement in Deerfield, Massachusetts, forty-seven settlers had been killed and more than a hundred taken captive. Deerfield was roughly forty miles upriver from Windsor, and those events undoubtedly made Connecticut colonists anxious about spending too much time in the woods.

There was yet another reason to engage in forest work that got you out of the woods quickly and without much capital investment: trespassing. Even for those with honorable intentions, it was easy to be on the wrong side of a township or colony boundary in the Connecticut River pine forests because there was significant disagreement about just where those lines were located. The problem dated back to 1642, when Nathaniel Woodward and Solomon Saffrey had been hired to survey the border between Massachusetts to the north and Connecticut and Rhode Island to the south. They began near the eastern (Rhode Island) end of the boundary, plotting a short line that closely followed the present-day border. One might have expected the pair to continue due west, drawing an unbroken line to the Connecticut River and beyond to New York. Instead, they had a bright idea. They would hop on a boat, pass along the southern coast of Rhode Island and Connecticut, and then sail up the Connecticut River to a point that was at the same latitude as their starting line on the Rhode Island–Massachusetts border. Then they would survey on either side of the Connecticut River to establish the line between Connecticut and Massachusetts. Wouldn't that be easier than bushwacking west from the northeastern corner of Rhode Island?[27]

Woodward and Saffrey almost got away with this. They made just one mistake, but it was a significant one. The surveyors marked a point along the Connecticut River that they thought was at the same latitude as the Rhode Island–Massachusetts line they had drawn at the beginning of their survey. But they goofed. The point they plotted was about eight miles too far south of the correct latitude; and as they extended the survey on either side of the river, it created a border that was also about eight miles too far south. As a result, when maps were drawn based on this survey, the Massachusetts Bay colony had bitten about 100,000 acres out of the top of Connecticut, meaning that land which should have been administered by Connecticut ended up being controlled by Massachusetts.

In the late 1600s, the Connecticut and Massachusetts governments filled in the map around the questionable border with newly incorporated townships. Leaders in Springfield, Massachusetts, including the merchant and major landholder John Pynchon, petitioned their colonial legislature to establish the townships of Suffield and Enfield in the area south of Springfield. Their goal was to protect their claims to the region's pitch pine forests, and the legislature agreed: both new towns were incorporated with southern borders that matched the Massachusetts view of the Woodward-Saffrey line. Meanwhile, landowners in Windsor, Connecticut, asked their own colonial legislature for a change in the borders of their township, originally established in 1637. They had their own ideas about survey lines and pushy Massachusetts colonists, and they wanted to move the town's original borders farther north to a line that would be north of the Massachusetts-favored boundary. The Connecticut legislature complied, also establishing the new town of Simsbury to the west of Windsor, with a northern border pressing up against Massachusetts's Suffield.[28]

Drawing lines and incorporating new townships was routine business for late-seventeenth-century colonial assemblies.

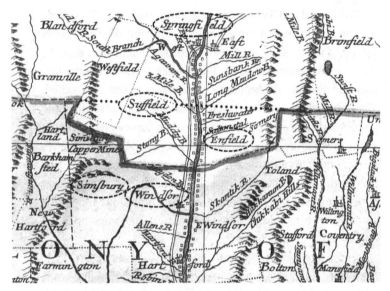

FIGURE 15. The Massachusetts Bay Colony takes a bite out of the Connecticut Colony. Suffield and Enfield were founded as Massachusetts townships. The dotted line shows the location of today's border, and relevant town names are circled with dashed lines. From Johann Michael Probst and Lukas Voch, *A Map of the Most Inhabited Part of New England* (Augsburg, Germany: Probst, 1779). Collection of the Library of Congress.

However, in the case of these river valley townships, uncertainty about the precise location of the Woodward-Saffrey survey line produced maps with overlapping colony borders. Depending on one's point of view, the northern borders of the Connecticut townships now ran north of the southern borders of Massachusetts townships or the southern borders of Massachusetts townships now ran south of the northern border of Connecticut townships. This created a narrow band of debatable land, about two miles wide from north to south, with the northern and southern borders near the mouths of Stony Brook and Kettle Brook, respectively. The ambiguity quickly led to turpentine-driven conflicts between township and colony governments.[29]

In the mid-1690s, valley residents increasingly turned from making tar in kilns to the lower-investment business of collecting raw turpentine. This transition was aided by the arrival of Huguenot refugees from France during the 1680s and 1690s. As Protestants, they were fleeing increased persecution from French Catholics, and many came from southwestern France, where locals had been gathering pine resin from the maritime pine (*Pinus pinaster*) for centuries. The Huguenots brought along their expertise in boxing trees and other aspects of turpentine production; and as the English colonists learned these new turpentining methods, output increased and border conflicts began to break out over control of the pitch pines.[30]

In 1699, disputes over naval stores pushed the citizens of Windsor to beg their colonial government to deal with alleged trespassers from Suffield and Enfield. These outsiders had "intrench[ed] . . . upon lands belonging to this Colonie, and by pattent, granted by this government to the town of Windzor." The Massachusetts Bay men had "forcibly enter[ed,] . . . cutting . . . timber, making tarre, and turpentine, and fencing and breaking up some of the said land, to the great damage of the said town of Windzor." In response, Connecticut's General Assembly ordered "that a committee be appointed, to treat with the government of the Province of the Massachusetts Bay . . . for a settlement of a line between this Colonie and said Province, that so this government, and the town of Windzor under them, may quietly possesse and enjoy what of right belongs to them, and they have so long enjoyed." Although this episode suggests that turpentine had become a commodity worth fighting over, the border issue remained unresolved for more than a decade.[31]

Meanwhile, events in Europe sharpened the conflict in the valley. The War of the Spanish Succession began in 1701, pitting a range of European countries against each other and pushing up the price of naval stores. Probably not coincidentally, the New England turpentine wars heated up in early 1703, when a Windsor constable arrested two Enfield men while they were turpentining. The men were taken to Hartford and brought before a justice of the peace for alleged crimes against Windsor. The turpentiners claimed they had been on Enfield land at the time of the arrest and took their grievances to the Massachusetts colonial council. According to the council's April 1703 minutes, the "Select Men of the Town of Enfield" complained "of the molestation and disturbance given them in the improvement of their estates by their neighbours of the Colony of Connecticot [sic]." The complaint stated that, on March 24, "the Constable of Windsor with six men, came near to the midst of their town bounds and seized the two men at work upon the turpentine." The Massachusetts government backed the Enfield men and drew up a letter to the governor and council of Connecticut. In it, they complained about the "irregular action" of the Windsor officer, arguing that Massachusetts had jurisdiction of the territory in which the arrests had occurred. They also declared that the arrested men "were commanded [by Massachusetts] not to appear at the [Hartford] court" and that the Massachusetts government expected "there would be no further procedure in that matter, and that the Queen's subjects be in peace."[32]

Despite this verbiage, Massachusetts towns took the offensive. In the fall of 1703, a Suffield posse arrested men from Simsbury who were gathering pine resin along the disputed border between the towns. The Simsbury men were charged with trespassing, and more than twenty barrels of turpentine were seized by the town of Suffield. Now it was Connecticut's turn to be outraged. The records of its colonial assembly note that "one Benjamin Dibble of Symsbury was about twentie dayes past arrested while he was about his lawfull labour within the town of Symsbury in this Colonie." The assembly viewed these proceedings as "very illegall and arbitrary, especially since it is found by observation and trial that the greater part of said Suffield is and ought to be part of this [Connecticut] Colonie." To no one's surprise, the colonial government concluded that Dibble's activities had been legal and had taken place on land that was or should have been controlled by Connecticut. Constables in Windsor and Simsbury were authorized to arrest "Jonathan Remington and Joseph Sheldin," the Suffield constables who had led the party that had detained Dibble. An eye for an eye, and an arrest for an arrest.[33]

Apparently, the Connecticut constables did manage to nab the two men as, in May 1704, the Massachusetts colonial assembly received a petition from "Joseph Shelden & Jonathan Remmington." The pair stated that, in late September 1703, they had "seized a parcel of turpentine Gotten by some of our Neighbours as Trespassers (within the settled bound of this plantation): And [we] were for our obedience to the authority of this province seized by the authority of Connecticot Colony, Carryed away to Hartford and thair committed to the common Goal [jail] where we suffered imprisonment for the space of three weeks and odd dayes." The petitioners hoped to be compensated for the losses they had incurred in defending "the authority of this province." In their view, they had fought for Massachusetts, and the colony owed them reparations. The cited the costs of posting bond to get out of jail and claimed they had lost critical time during planting season, suffering "exceeding great Damage it being in the spring and in the very prime and aptest season for sowing and planting soe that thereby we are like to sustain the loss of this years Crop . . . [as a result of] our exilement from our families and imprisonment at Hartford."[34]

It took time for governmental wheels to turn; but in November 1705, the Massachusetts assembly resolved to pay £10½ to Shelden and £8½ to Remington. Shelden received a little more because he'd spent extra money on the petition. The resolution sourly noted that the compensation was intended to cover "their charges and damages by being imprisoned at Hartford for asserting the right of this government to the said township against the encroachments of the government of Connecticut." Shelden and Remington had upheld the honor of Massachusetts, and the assembly was making sure that Connecticut knew it.[35]

As the two colonies were trading barbs and arrests, Parliament passed the Naval Stores Act of 1705, which gave settlers in the colonial borderlands even more reason to care about their boundary lines. The act had been prompted, in part, by a conifer species distribution issue similar to the one that had plagued the Royal Navy with respect to ships' masts and white pine (see chapter 2). Without any remaining pines of its own, Britain was now importing most of its naval stores from Scandinavia, controlled by the Swedish Empire. While the Swedes were usually willing to sell, Britain could not take this for granted. Even when shipments were flowing smoothly, the price was controlled by Swedish merchants who had monopolized naval stores under the banner of the Stockholm Tar Company. During the 1690s alone, the company had doubled the price of barrels of tar and pitch. When England joined the War of the Spanish Succession (which would continue until 1713), the Royal Navy's demand for pine products rose, and the price doubled again. In early 1700s, when Sweden became involved in the Great Northern War (1700–21) against Russia and its allies, there were further disruptions to the flow of tar, pitch, turpentine, and rosin.[36]

British officials were well aware that the American colonies could be a source for naval stores—a source that would be controllable at every step, from the pine trees to the ships' rigging. Beyond its military importance, total access to American pines would support the goals of the British Empire's expanding mercantile policies, designed to create a self-sufficient economic system with complementary parts. The British Board of Trade and Plantations had been established in 1696 to promote these policies, which included reducing the outflow of gold and hard currency to other nations, increasing trade between the colonies and Britain, and discouraging the development of colonial industries that might compete with industries at home. Buying tar and turpentine from the Americans could break the Swedish monopoly and lower the expense of naval stores, thus accomplishing two goals at once: reducing English dependency on foreign sources and decreasing the outflow of gold. The colonies would justify their existence by sending raw, pine-based materials to the mother country, and this would give them enough income to purchase manufactured goods from England. And if the colonists were profiting from naval stores production, they would have less need or incentive to develop their own manufacturing capacity. In the opinion of the trade commissioners, North America's *Pinus* species could be the solution to a long list of England's problems.[37]

But there was a hitch in this plan. Though New Englanders had eagerly chopped down white pines for masts, they weren't that excited about producing naval stores. As I've noted, not only were labor costs were high, but the colonists had other uses for pitch pines and could earn money in many other ways. As a result, between 1701 and 1704, American tar and pitch accounted for no more than 2 percent of the amount imported into England, while Sweden continued to supply 70 to 85 percent of Britain's needs.[38]

What the colonists needed was a reason to care more about naval stores. The Crown had employed legal sticks in an effort to slow the cutting of white pines, but it would take some type of carrot to spur them to harvest more tar and turpentine. So the Board of Trade recommended adding a premium to the price that England paid for each barrel of naval stores. The Naval Stores Act of 1705 turned this recommendation into law. It declared that the Royal Navy and the "wealth, safety, and strength of this kingdom" depended on the "the due supply of stores necessary for the same." Unfortunately, these stores were "now bought . . . from foreign parts, in foreign shipping, at exorbitant and arbitrary rates." Yet the American colonies had been specifically "design[ed] to render them as useful as may be to England." So now, "if due encouragement be given," these colonies would have a mandate to supply "great quantities of all sorts of naval stores." The "encouragement" would take the form of a bounty of £4 per ton of tar or pitch and £3 per ton of turpentine or rosin. It's likely that the premium was higher for tar and pitch because

England's fleets consumed considerably more of these stores, leading officials in London to focus on that output. As a bonus, the act would "further [the] imployment [*sic*] and increase of English shipping, . . . [and] the trade and vent of the woolen and other manufactures and commodities of this kingdom . . . in exchange for such naval stores, . . . enabling her Majesty's subjects, in the said colonies and plantations, to continues to make due and sufficient returns in the course of their trade."[39]

After passage of the Naval Stores Act, colonial exports of tar and pitch increased from an average of 450 barrels in 1701–4 to an average of 7,350 barrels in 1706–9.[40] In the latter period, American tar and pitch accounted for about 20 percent of English imports of these commodities, and the percentage would continue to rise in the coming decades. Although the bounty for raw turpentine was lower, it, too, increased in economic value and gave the settlers of the Connecticut River valley one more reason to tussle over borders and pine trees.

In 1708, officials in the Massachusetts towns of Enfield and Suffield continued to seize turpentine and arrest Connecticut harvesters. Likewise, in Windsor, justices of the peace issued writs for the arrest of Massachusetts trespassers, and at least one Enfield resident was detained in March. In early April, the turpentine tempest reached its riotous climax after a Windsor constable and his deputies went into the woods intending to seize Enfield harvesters on Windsor-claimed land. Although the targets managed to evade arrest, they were angry; and with a group of their neighbors, they began hatching plans for a fight. It's uncertain how many men on both sides were drawn into the subsequent rumble. However, years later, Connecticut's governor, Roger Wolcott, himself a Windsor native, recalled that "the inhabitants were So fiercely Engaged that I [saw] . . . Near a Hundred men Meet to Decide this Controvercy, by force, [and] a Resolute Combat Ensewed between them in which many blows were Given to the [exasperation of] Each party. So that the Lives and Limbs of his Majesty's Subjects were Endangered thereby."[41]

By the time the fighting was over, the Windsor sheriff had taken five Enfield men into custody. They were brought to a court in Hartford and held for trial, with bail was set at the dizzying sum of £100. However, before the trial could take place, government officials in both colonies concluded that the turpentine quarrel had gone too far. All parties agreed that they needed a permanent solution to the border question. Thus, to improve the climate for negotiations, the Connecticut courts dropped the charges against the Enfield men.[42]

In 1713, the colonies reached a somewhat convoluted settlement about the boundary. Massachusetts agreed that the entire length of the border with Connecticut would follow the same latitude line as the Massachusetts–Rhode Island border without significant deviations. This clearly put Suffield and Enfield south of the border, making them part of Connecticut. Yet Massachusetts would retain jurisdiction over these towns because, early in the negotiations, officials had decided that towns would remain with the governments under which they had been settled. Perhaps this was to avoid invalidating the land grants that Massachusetts had made to the colonists of Enfield and Suffield. Regardless, the decision subtracted about 110,000 acres (about 170 square miles) from Connecticut's administrative domain. As compensation, Connecticut received rights to equivalent lands controlled by Massachusetts. The colony quickly auctioned off this newly acquired land and used the money to provide funds for its new Collegiate School in New Haven, better known today as Yale University.[43]

The settlement also nailed down the borders between Windsor to the south and Enfield and Suffield to the north. The boundaries were drawn south of the line that Windsor had claimed during the border controversy; as a consequence, about 7,260 acres of disputed land (a little more than eleven square miles) was awarded to Enfield and Suffield. Windsor was compensated with equivalent lands taken from the township of Tolland located on Windsor's eastern border. (How the people of Tolland felt about this arrangement is not known.) The final lines may not have satisfied everyone, but at least those who lived in the Connecticut River valley now knew whose pine trees were whose. Decades later, in 1749, Connecticut would annex Enfield and Suffield at the request of these townships, leaving us with the clear Massachusetts-Connecticut border we have today.

Turpentine had forced the resolution of a festering problem between two colonies and had inadvertently funded the early development of Yale University. But by the time the border dispute was settled, the issue of pitch pine forest ownership didn't much matter anymore. By the 1710s, the New England tar and turpentine boom was ending. After so much extraction, the forests had significantly reduced in size, and northern colonists could not compete with the fast-growing naval stores industry in North Carolina. That colony's sandy coastal plains had little agricultural potential, but they supported endless forests of loblolly pine and yellow, or longleaf, pine, which were ideal for producing resin. In addition, the region had access to cheap labor—that is, to enslaved workers.

By 1720, the American colonies as whole were supplying 80 to 90 percent of England's tar imports, although the Royal Navy still preferred the Scandinavian product. The Finns and Swedes consistently produced higher-quality tar, while the Americans preferred to take shortcuts, making a debris-filled product from dead wood and freshly cut trees. The colonists were not going to wait three years for a trunk to get gooey. Yet regardless of the quality issue, New England was finished as a significant producer of naval stores. Going forward, North Carolina would generate the vast majority of North American tar and turpentine, eventually becoming known as the Tar Heel State.[44]

Pine tar played one other role in the traditions of eighteenth-century New England. It was the adhesive of choice for tarring and feathering. In this ritual, the tar could be applied cold, or it could be heated to make it easier to spread, which risked causing severe burns. If victims were lucky, most of the tar went onto their clothing, but sometimes they were was partially undressed to increase the amount of skin coated with adhesive. After the tar was applied, feathers from beds and pillows were dumped onto the sticky skin and clothes. Then the sufferer might be forced to parade through town in a cart or straddled over a fence rail, a practice known as "riding the rail." Removing tar from skin was painful. At a minimum, it involved scrubbing with a solvent such as turpentine, and the process could tear away strips of skin, creating a significant risk of deadly infections in an era without antibiotics. The burns from hot tar created a further threat of scarring and infection. The vast majority of victims survived, but the ordeal could be fatal.[45]

The practice of tarring and feathering was already centuries old when the American colonists first adopted it as a political tool in their response to the Townshend Revenue Acts of 1767, which taxed a wide range of items imported into the colonies, including glass, lead, paint, paper, and tea. The acts, which were designed to raise revenue to support the empire's presence in North America, also asserted the British government's authority to collect money in this manner. Although the monies were to be used exclusively within the colonies, enforcement would require officials to staff a Boston-based Board of Customs and employ a network of informants to identify tax evaders.[46]

The Townshend Acts sparked strong opposition. Americans considered them to be taxation without representation, for the colonists had no formal voice in the British Parliament. To show their displeasure, the Sons of Liberty, a loose political organization founded in about 1765, targeted both customs officers and informants for tarring and feathering, often with the aid of other less well-organized citizen supporters. In 1768 and 1769, men were tarred and feathered in the Massachusetts towns of Salem, Gloucester, and Boston as well as in Philadelphia and New York City. In one case, an innovative crowd in Salem applied the feathers by repeatedly tossing a live goose at a freshly tarred man named John Row (or Rowe), a customs official assigned to board merchant ships and secure payments.[47]

The Sons of Liberty also organized boycotts in which colonial merchants agreed not to import British goods. However, these boycotts cut into sales, so some shop owners quietly set them aside. In the spring and summer of 1770, the Sons used a variation on a sticky theme to try to persuade Boston backsliders to stand firm on the boycotts: instead of tarring and feathering the merchants, they tarred and feathered their houses and stores. We don't know exactly why they resorted to this gentler approach. Possibly the Sons recognized that the merchants were only indirectly involved in collecting Townshend duties. Or perhaps some of the Sons themselves were quietly selling British goods, and they may have felt uncomfortable about attacking others for doing the same thing. Regardless, the boycott could not be sustained; and in the fall of 1770, Boston merchants voted to discard all nonimportation agreements. It helped those opposed to the boycott that some of the steam had been taken out of the issue by a partial repeal of the Townshend Acts in April 1770.[48]

In April 1770, the partial repeal led to a period of relative calm, but this ended when the Tea Act of 1773 imposed new duties on imported British tea. Among the American colonists, the act sparked a renewed emphasis on both boycotts and tarring and feathering and, in December 1773, led to the famous Boston Tea Party. When word of the Tea Party reached England, an agitated Parliament passed the Coercive Acts of 1774. These, in turn, spurred colonists' formation of the First Continental Congress in the fall of that year. During meetings, the delegates managed to agree that the colonies should boycott all British goods. To enforce nonimportation, it promoted the formation of local committees that would identify those who did not comply, and many of those committees would use tarring and feathering as a means of reprimand. Later, the practice would be expanded to punish those who expressed disapproval with the congress or the ongoing American Revolution, and before long the practice became as firmly associated with the era as "Yankee Doodle" was.[49]

Tarring and feathering would continue into the 1800s, and those who mentioned the practice knew that an audience would understand what they were saying. Thus, in the 1860s, Abraham Lincoln, after being asked about how it felt to be president, said that he was reminded of a story about a man who was tarred and feathered and ridden out of town on a rail. When asked about the experience, the man replied: if it were not for the honor of the thing, he would much rather have walked.[50]

Chapter 8

Hemlock Tannins and Making Leather

IN 1872, DELEGATES AT the Republican Party's convention nominated the incumbent president, Ulysses S. Grant, as their candidate in that year's election. On the surface, he seemed like a shoo-in: not only was he currently in office, but he was also a Civil War hero. Still, the party was not taking any chances. Hoping to attract more working-class voters, Republicans created a campaign poster that depicted Grant's running mate, Henry Wilson, in the garb of a "Natick Shoemaker" and Grant himself as the "Galena Tanner," dressed in a tanner's apron and standing in front of a tannery door. Never mind the siege of Vicksburg and the destruction of the Army of Northern Virginia, Grant was a man who knew how to make leather.

Grant went on to win almost 56 percent of the popular vote and 286 out of 352 electoral votes, so it seems unlikely that it was necessary to hang an apron around Grant's neck to win the 1872 election, but the campaign poster did contain a germ of truth: as a young man, Grant had toiled in his father's small tannery in Georgetown, Ohio. In his memoirs, the president recalled: "While my father carried on the manufacture of leather and worked at the trade himself, . . . I detested the trade, preferring almost any other labor."[1] Before the war, Grant turned instead to farming, but his failure in that venture forced him to return to working for his father, who by then owned a leather-goods store in Galena, Illinois. Years later, Grant's memories of the leather trade remained bitter.

FIGURE 16. Ulysses S. Grant depicted as a tanner on a presidential campaign poster, 1872. Collection of the Library of Congress.

As both an occupation and technology, leather making stretches back for thousands of years. Though there are variations in approach, the practice has always depended on a group of plant-derived molecules called tannins. A wide range of plant species produce tannins; but to turn animal hides into leather on a commercial scale, tanners need trees that produce high concentrations of them. In nineteenth-century America, oaks were the most important source in the central and southern United States, and Grant's father probably used oak-based tannins in his work. However, in New York and in much of New England, the eastern hemlock was the foundation of the industry. Like other conifers, hemlocks make and store plenty of terpenes in their sap and needles, but they also produce high levels of tannin, especially in the bark, which may be up to 12 percent tannin.

The tannins so essential to leather production are relatively large, water-soluble molecules that can bind to proteins and link them together. In hemlocks, the tannins are manufactured by combining molecules that contain three rings of carbon atoms into much larger molecules that have dozens of rings of carbon. Attached to these rings are many atoms of oxygen and hydrogen, also called hydroxyl (–OH) groups. These abundant hydroxyl groups stick directly to proteins, allowing a tannin molecule to link together many protein molecules, a key reaction in leather making.[2]

Certainly, hemlocks are not interested in our desire to make leather. So why do they make tannins? There's a cost to creating

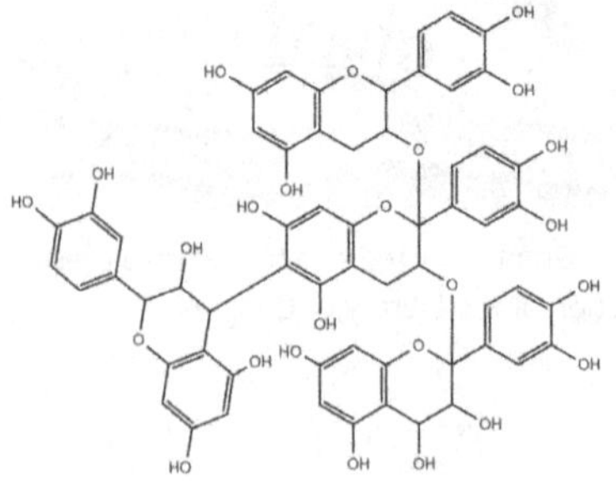

FIGURE 17. A tannin composed of rings of carbon. (Though not shown here, there are carbon atoms at any point where two lines meet.) Note the many –OH groups attached to the rings.

these molecules because carbon and energy put into tannin production are resources diverted from growth and reproduction. That means tannin creation must somehow benefit the tree, likely as a way to protect itself against herbivores and disease-causing microbes. Hemlocks can live for more than five hundred years and over that vast span of time are bound to be targeted by pathogens and plant eaters. It makes sense that such trees would produce high concentrations of defensive chemicals such as tannins and terpenes.

Tannins protect hemlocks from herbivores in ways that are closely related to how tannins tan leather: both involve the linking of tannin molecules and proteins. Some ecologists suggest that tannins discourage hungry insects by clumping together plant proteins in their digestive tracts, making it difficult for digestive enzymes to break down the proteins into amino acids. Because the gut enzymes themselves are proteins, the tannins might also directly bind to and inhibit them. If the insects can't split large proteins into small absorbable amino acids, then they will suffer from amino acid deficiencies and malnutrition.[3]

Of course, many plant-chewing insects have evolved a strategy to counter tannin's protein-binding abilities. Some have very alkaline (high pH) midgut sections in their digestive tracts; in this environment, tannins don't stick well to proteins so are less likely to interfere with protein digestion. Still, such digestive tracts may be vulnerable to damage from a different tannin effect. When an insect gut is very alkaline, tannins undergo chemical changes that create highly reactive molecules. These damage digestive tracts by oxidizing vital biological molecules and cell components. So, for some bugs, tannins may work as poisons rather than digestion inhibitors.[4]

Mammalian herbivores do not have highly alkaline digestive tracts. Instead, their guts often have a very acidic chamber that is comparable in pH to the human stomach. Tannins bind well to proteins at acidic and near neutral pH levels, and there is evidence that they do harm some mammalian herbivores by interfering with the digestion of proteins. Thus, many foliage, twig, and bark-chewing mammals have evolved countermeasures. Some species produce saliva proteins containing a high percentage of the amino acid proline. Proline-rich proteins (PRPs) tend to be long and stringy, with many tannin-binding atoms on their surfaces. When a mammal produces PRPs, they become entangled with any tannins in its food. The proteins coil around the tannins, forming compact blobs; and the tannins form bridges between separate PRP molecules, creating tiny particles composed of tannins and linked proteins. In the way, the PRPs inactivate the tannins, essentially removing them from the food so that they cannot interfere with protein digestion farther down the digestive tract. There is a metabolic cost to producing PRPs, but they are also an effective way to remove troublesome tannins.[5]

Mammals vary in their ability to produce PRPs. However, animals such as foliage-browsing deer tend to produce larger quantities than do species with low-tannin diets, such as grass-eating sheep. Humans are among the mammal species capable of producing PRPs, and you can experience the result when you consume a high-tannin beverage such as red wine. Tannin-rich items have a property called *astringency*: they create a rough, dry, puckering sensation in the mouth, which takes place when the tannins bind to PRPs and other saliva proteins. In fact, the term is derived from the Latin *ad stringere*, meaning "to bind." The sequence of events goes something like this. When we consume tannins, they initially bind to the abundant PRPs in our saliva. This reduces saliva's ability to provide lubrication, and the rough sensation in our mouths may also be linked to the precipitation of tiny tannin-protein particles onto its inner surfaces. If we happen to consume more tannins than the PRPs can sponge up, the excess can trigger additional reactions.[6] That's because tannins also bind to viscous proteins called mucins. These are bound to epithelial cells on oral surfaces, where they help to form a thin, slippery film called a mucosal pellicle, which reduces friction and abrasion between the tongue and other mouth parts. However, when tannins combine with mucins, the proteins begin to clump together, and the pellicle changes from a thin, slick film to a bumpier layer that creates more friction and less lubrication. Mucin clumping may also expose taste receptors in the epithelium beneath the pellicle. There is evidence that some tannins may bind to receptors that detect molecules that we perceive as bitter, which accounts for why some people some consider bitterness to be part of astringency.[7]

Regardless of the exact mechanism, tannins make us pucker. In red wine, their presence is part of the wine's body, or structure, although reactions to a wine's astringency vary with the individual. Tannins can be detected by many types of herbivores, and some species and individuals appear to find the astringency sensation so unpleasant that they refuse to consume certain tannin-rich foods or plants. Our reactions are truly a matter of taste.

What does all this have to do with hemlocks and tanning leather? The chemical changes that take place in animal hides during leather making mimic what happens between tannins and PRPs in the mouth. In the case of leather, the main protein of interest is collagen, a long, stringy protein loaded with proline. Strands of collagen can wrap around each other like threads in a rope, creating strong, fibrous materials that are an important and abundant component of tendons, ligaments, and skin. This protein is also ideal for binding tannins. Like salivary PRPs, collagen has lots of tannin-binding atoms on its surface, allowing them to easily interact with the tannins' hydroxyl groups. During leather making, the tannins soak through the skins, attach to the collagen, and form bridges or cross-links between separate collagen fibers. This makes the material stronger, more durable, less likely to decompose, and more water-resistant.[8] Tannins also change the color of skins: oak tannins dye them yellowish tan, while hemlock tannins produce dark brown. The preservative transformations caused by tanning have been known for centuries. For example, in *Hamlet*, written in about 1600, William Shakespeare included dialogue about tanning that playgoers would have easily understood. When Hamlet asks the gravedigger,

"How long will a man lie i' the earth ere he rot?" the gravedigger explains that a tanner's body takes longer because "his hide is so tanned with his trade, that he will keep out water a great while; and your water is a sore decayer of your whoreson dead body."[9]

The temperate forests of North America once contained vast stands of oak and hemlock, and the history of the continent's tanning industry dates back to the 1630s, when Dutch settlers operated large tanneries in New Amsterdam (now New York City). By the early 1800s, the general procedures for tanning leather in North America had been well established. Workers began the process by soaking hides in water to remove flesh and fats and then in highly alkaline lime solutions to loosen or dissolve hair; this also caused the hides to swell and thus open bundles of collagen fibers for later penetration by tannin solutions. Next they neutralized the lime solution with vinegar and spread the hide over a beam. In what was known as beaming, workers scraped off any remaining hair and used a sharp knife called a flesher to carved away remnants of flesh and fat. After rinsing the hides in water, tanners scraped the interior and exterior layers with a curved blade to produce a smooth surface with uniform thickness. At this stage, hides were often cut into two separate pieces, known as *sides*. The material was now ready for the main event: soaking in tannin solutions to create leather.[10]

In most tanning operations in New York and New England, leather makers extracted tannins by shredding or grinding kiln-dried hemlock bark in a mill to produce a coarse powder that was between 10 and 12 percent tannin. Workers added this powder to tubs of hot water as if they were brewing huge cups of bark tea. It could take them several days to extract the tannins, although they could shorten that time by infusing the bark with steam or diluted sulfuric acid. Aptly, the tanners referred to this pungent product as *ooze*.

Prepared hides were immersed in a series of vats containing increasingly concentrated tannins. It was critical to begin with a diluted solution; otherwise, excess tannin would coat the outermost layer of collagen and create a stiff, repellant crust that

FIGURE 18. The beaming stage of leather production. From Charles Thomas David, *The Manufacture of Leather* (Philadelphia: Baird and Company, 1897), 131.

would eventually prevent tannin molecules from reaching the hide's interior layers. In other words, starting with a high concentration of tannin would be like sandwiching an internal filling of untanned skin between two slices of tannin-rich bread. In the earlier stages of immersion, the hides might be tossed into vats and churned with plungers, or they might be hung on racks that were then dipped into the vats. In later stages, racks of hides might be dipped into stronger tannin solutions, or the hides might be flattened and stacked in empty vats, with shredded or ground bark sprinkled between each layer. This method not only increased the tannin level but also added spaces between each hide into which the tannin extract could flow when the vat was subsequently filled with ooze. In the end, the hides would have soaked in about a half-dozen vats, a process could take several months to more than a year to complete, depending on what the final product would be used for. For instance, leather intended for shoe soles had to be thicker and heavier than leather for the uppers and thus required more time in the tanning vats. At the end of the soaking period, tanners would drain the ooze from the hides and hang them in a shed or a loft to dry.

During the first half of the nineteenth century, small-scale tanneries were in operation throughout the United States. The 1840 census recorded thousands, most of them employing a just a few workers each. However, as demand for leather grew, tanneries expanded in size. These larger factories consumed huge quantities of tanbark, and the New England–based industry required extensive hemlock forests to support this growth. Where could tanners find the habitats that would support such forests?

Hemlocks are usually associated with acidic soils that are sandy or silty and filled with coarse rocky material. In New England and southern Canada, such soils have accumulated over the course of centuries on the till left behind by retreating glaciers. Hemlock-favoring soils are also usually moist but with good drainage and are often located in coves, at the base of ridges, and along streams, lakes, and bogs. Yet, paradoxically, hemlocks may also grow on drier rocky slopes and ledges, possibly because they are supported by subsurface seeps of water. Once a habitat has been colonized by hemlocks, falling and decomposing needles increase the soil acidity, further benefiting the trees.

The seeds of hemlocks often germinate on surfaces with good water retention and a nutrient source, such as damp and rotting logs and stumps with a light moss cover. By starting life on logs and stumps, these tiny plants are less likely to be buried in falling leaves. Because the seedlings require moisture and partial shade to become established, they do not compete well with fast-growing, dry soil–tolerant species such as birches, aspens, red maples, and pines that pioneer in open habitats created by farming, logging, or fire. But as a forest matures, the opportunities for new hemlocks increase over time, as canopy shadows keep the ground damp, decaying logs and stumps become more common, and soils become more acidic due to decaying conifer needles. For most tree species, starting life under dim light would be a formula for failure, but hemlocks are remarkably shade-tolerant, and seedlings can survive in as little as 5 percent of full sunlight. Probably not coincidentally, they grow more slowly than most other North American conifers.

Very high shade tolerance and a long lifespan make hemlocks well adapted to conditions in mature or late-succession forests. In favorable habitats, trees can reach eighty feet in height and about a foot and a half in diameter after a hundred years, more than a hundred feet in height with two- to four-foot diameters after two to three hundred years, and up to six feet in diameter at five to six hundred years. With time, hemlocks produce old-growth forests in which they are either the dominant species or co-dominant with other long-lived, shade-tolerant species. Large trees create microhabitats of deep shade and cool, moist, acidic soils that favor the development of a hemlock understory and the perpetuation of the hemlock woodland. But a low capacity for colonizing open habitats, a slow growth rate, and a tendency to adapt to conditions found in older forests mean that, once an old-growth hemlock forest is removed by logging or fire, it can take a long time to regenerate. This had significant consequences for the nineteenth-century tanning industry.

Maine contains millions of acres of habitat where conditions are favorable for the development of old-growth hemlock forests, which is why the state became a major leather producer in the 1800s. According to *The Tenth Annual Report of the Bureau of Industrial and Labor Statistics for the State of Maine* (1896), Maine had almost four hundred tanneries in 1840, most of them located in the south-central and western regions and employing an average of two workers per operation. In the years just before the Civil War, the total number of tanneries had dropped to about 150, but their scale had expanded, with each employing an average of about five workers. Then, after the war, their size increased dramatically. Although hemlock bark had become scarce in the older, more settled counties to the south and west, massive tanneries were being established in the eastern counties between the Penobscot and Saint Croix rivers, where coarse, acidic, boggy soils in a land dominated by numerous streams, ponds, and lakes had encouraged the development of extensive old-growth hemlock forests. By 1880, Maine was down to eighty-five tanneries, but the average number of workers had increased to about twenty per operation, with some facilities employing as many as a hundred wage earners. Often several tanneries would be under the control of a single ownership group.[11]

In Maine, the Shaw brothers came to dominate leather production. They were the sons of Brackley Shaw Sr., who had begun tanning leather in the 1820s. By the 1840s, the senior Shaw was well established in Cummington, Massachusetts, in the hemlock-rich Berkshire Mountains, but he lost much of his business in a fire and then died of a fever in 1848. His sons Fayette, William, and Thackster and their uncle, Charles Shaw, rebuilt the business in the 1850s with a plan to put various family members in charge of the different locations of the operation. Soon the boys were directing an expanding collection of family tanneries in central Maine's Sebasticook Valley, in towns such as Detroit, Plymouth, Burnham, and Dexter. By the start of the Civil War, the family's Maine operations accounted for about half of all of the sole leather produced in Maine. But the Shaw boys were just getting warmed up. After the war, the family expanded its business into eastern Maine, building several large tanneries near the Canadian border in Kingman, Jackson Brook, Forest City, Vanceboro, and Grand Lake Stream.[12]

The Grand Lake Stream facility would be the Shaws' magnum opus. The brothers built it in the 1870s at an ideal location on a southern extension of Grand Lake (now called West Grand Lake). Extensive hemlock forests surrounded the site, and a chain of upstream lakes provided a water route for the transport of hemlock bark to the tannery. At the facility, the bark was floated to a mill for grinding bark via a 350-yard-long canal excavated along the east bank of the stream draining Grand Lake. A leach house sat next to the bark mill, where workers made tannin-rich ooze. A furnace adjacent to the leach house heated water for extracting the tannin from the bark and, to close the circle, it was fueled by bark waste from the leach house. The furnace's impressive eighty-foot-tall smokestack stood as monument to the tannery well into the 1900s, long after most of the rest of the abandoned

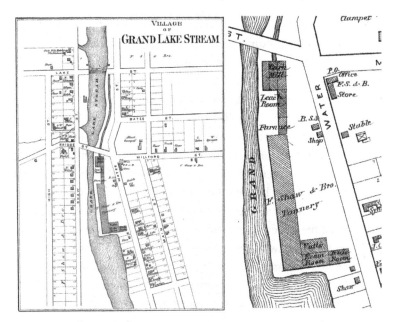

FIGURE 19. The village of Grand Lake Stream in the 1880s. Left-hand map: The tannery is on the eastern bank of the stream, with a canal leading to the site running to the north. Right-hand map: The bark mill and the leach room are at the northern end, the tanyard (tannery) building with hundreds of vats is in the middle, and the beam room and the hide room are at the southern end. From *Atlas of Washington County* (Houlton, ME: Colby, 1881). Collection of the Maine State Archives.

facility had been dismantled or burned. A ten-story tower, housing a drying loft, had a bell in the cupola. It summoned workers to the tannery at 6 a.m., announced a brief recess at noon, and dismissed the crew at 6 p.m. In July 1885, the bell tolled to mark the death of the Galena Tanner, Ulysses S. Grant.

Raw hides were shipped in from as far away as South America, thousands of miles to the south. Tanneries needed about ten tons of bark to produce one ton of leather, so it made sense for the hides to come to the hemlocks instead of the reverse, despite the distance they had to travel. Before the ooze could work its magic, the hides had to be properly prepared; and in Grand Lake Stream, tanners relied on methods that used more physical pummeling and microbial decomposition than had been typical in earlier nineteenth-century leather making. In the beam room, workers soaked the sides in water for several days and then sent them to a hide mill, where another set of employees pounded the skins with timber beams to soften them and increase their pliability. Afterward, the milled hides were taken to sweat houses— underground vaults in which the sides were hung up and left to partially decay for several days. The heat from decomposition and the ammonia generated from the breakdown of proteins loosened the remaining hair and flesh, but workers had to take care, for too much heat and decay could damage the hides. And the stench must have been strong enough to peel the skin off the workers as well. The sweating process was followed by another round of hammering at the hide mill, and then the materials returned to the beam room, where beamsters used fleshers and beam knives to remove any remnants of hair and flesh. The use of hide mills reduced the amount of scraping required, but beamsters were still prone to losing fingernails, and many suffered from festering sores when the scraping of hides rubbed forearm skin raw and opened the arms to infection.[13]

The output from the beam room, known as green leather, was sent to the tannery's eighty- by six-hundred-foot-long tanyard building. In her history of Grand Stream Lake, Minnie Atkinson described the tanning floor, claiming that it contained six hundred tannin-filled vats arranged in ten rows of sixty each. Another account, by Amos Wilder, stated that there were seven hundred vats.[14] Either way, there were a lot of vats. Each held up to 150 sides, and these were added one at a time, with a small quantity of leached bark sprinkled between each layer to give the ooze free access to all surfaces. Tanning sole leather required soaking in five or six different liquors of increasing tannin concentration, and the Grand Lake Stream tannery had rails laid inside to help move the sides from one vat to the next. As the ooze's tannin concentration intensified, the tanners increased the soaking time in each vat because the hide's absorption of tannins decreased over time as the material became impregnated with the chemicals. Sides might spend four to seven days in the low tannin solution of the first vat but up to forty days in the last vat's highest concentration. With sole leather, the goal was to give the material all the tannin it could absorb to create a firm and thoroughly preserved product. Additional tanning also increased the product's weight; and because sole leather was sold by the pound or ton, leather makers had an incentive to insist on a long tanning time. By contrast, leather destined for the uppers was sold by the square foot. It required less tannin per side and could be made where the bark was more expensive.[15]

After spending as long as six months in the vats, the sides were removed, and the excess tan liquor was rinsed off. The new leather was coated with cod oil and hung up to dry for one or two weeks in Grand Lake Stream's drying loft, at the top of a the company's tower. Then the leather was taken to the rolling loft, where it was dampened with water and pressed with a metal roller to compact the fibers and lay the grain. After several rounds of rolling, the smooth glossy leather was thoroughly dried, stamped with its weight and the tannery's name, and shipped to markets.

At its peak in the early 1880s, the Grand Lake Stream facility was probably the biggest tannery in the world; and the writer Amos Wilder, already mentioned, shared a glimpse of the massive operation in the June 2, 1883, issue of the Kennebec Journal. Wilder (who would become the grandfather of the playwright Thornton Wilder) had gone to Grand Lake Stream on a fishing trip and was disturbed to note that the tannery had negatively impacted the region's "land-locked salmon." As he wrote, "were it not for the government [fish] hatching houses . . . the taking of these game fish . . . would soon be known only as pleasure of the past." Nonetheless, he apparently had an interest in tanning facilities, for he wrote that there were more vats at Grand Lake Stream than at any other Shaw tannery in Maine and that the facility appeared to be running at maximum capacity: "The tannery runs day and night the year round (Sundays excepted) and turns out one thousand sides of leather every working day." Around-the-clock operations were possible because the "great tannery is lighted with Edison's latest and most approved incandescent electric light." Grand Lake Stream was at the cutting edge: it had electric lighting before the city of Bangor did, and "next month the great inventor of this light is to make this place a visit, and should this light prove a success in this tannery it will be introduced into all the others that are running." If Wilder was correct and the tannery was indeed operating at full capacity, six days a week, then it alone would have been producing about twice as much leather per year as were all of the Shaws' other leather operations in pre–Civil War Maine combined.[16]

Grand Lake Stream may have been a great place to tan leather. However, because the facility had to be located near relatively untouched hemlock forests, there were no nearby established towns to house the workers. This meant that the Shaw brothers had to build their own company town in this remote region, with room for about one hundred full-time employees and their family members. The new village featured tenement houses, hotels, a schoolhouse, and a company store that supported a total population of up to four hundred inhabitants. But it was not exactly a worker's paradise. The town was littered with boulders, stumps, and holes where stones had been pried out of the ground for foundation walls. Streets were rough tracks for tannery horse teams, and the houses were unpainted inside and outside. As was usual with tanneries, industrial waste polluted the air and the streams. On the bright side, Amos Wilder reported that there was no tax on dogs.[17]

One of the Grand Lake Stream workers, Martin Butler, would go on to a career as a poet, a printer, and a newspaper editor. He recorded his impressions of the company town in his poem "The City of Grand Lake Stream," published in his 1889 collection Maple Leaves and Hemlock Branches. Butler described the streets as "narrow, dark and drear," running "o'er rocks and stumps we have to jump." Workers earned "very scanty pay" for " grinding bark, or scraping on the beem [sic]," and they spent most of that money at the Shaw-controlled company store. During his years in Grand Lake Stream, Butler had experienced firsthand the dangers of the tanning industry, including an accident in the 1870s in which his arm was caught and mangled in a bark-grinding machine. Though his arm had to be amputated, he received no compensation from company headquarters in Boston. But industrial accidents and company indifference were the norm in the nineteenth century.[18]

The tannery was located close to extensive stands of hemlock, but the bark still had to be harvested. An operation the size of Grand Lake Stream's might require two or three hundred men to provide enough bark to meet demand. Cutting pine and spruce trees was a winter occupation, but the hemlock bark harvest, or peeling season, ran "from the full moon in May to the full moon in August." That's because it was easier to peel bark in the summer, when sap flow was at its peak. Since the weather was warm, there was no need to build sheltering camps, but the woodsmen had to endure swarms of Maine's infamous mosquitoes and blackflies, the "defenders of the wilderness." To save themselves from the insect onslaught, harvesters rubbed a thick mixture of pine tar and lard over any exposed skin, but their days still must have been miserable. As Butler wrote,

> 'Tis then, those bark-peelers will rage
> And tear their hair and scream
> And the curse the day they ever came
> To the City of Grand Lake Stream.

No doubt the one-armed poet had his own reasons to join to that communal curse.[19]

Peeling crews were usually composed of four workers: a chopper, a knotter, a ring-and-splitter, and a spudder. Once the team had selected a hemlock, the ring-and-splitter would cut one ring around the base of the tree and another about four feet above the first. Then he would split the bark in a line connecting one ring to the other. This seam gave the spudder a place to start peeling. To do so, he would insert a spud between the bark and the tree's sapwood. The spud looked a bit like an oversized carpenter's chisel; it had a two-and-a-half-inch blade, rounded on the edges and curved slightly downward so it could follow the convex surface of the trunk. After the spudder had removed this first four-foot-long section of bark from the base, the chopper would cut down the tree, and the knotter would remove the branches along the length of the trunk. The ring-and-splitter would then divide the bark into more four-foot sections, and the spudder would follow behind.

During peeling season, the workers left piles of bark throughout the hemlock stands. Then, in the fall, three-person yarding crews would move those piles to yards located along rough forest roads. First, a swamper would cut a narrow path winding among logs and brush to reach the small piles of bark. Then a teamster would drive in horse teams attached to jumpers—small sleds with hardwood runners—that could each carry about half a cord of peeled bark. These teamsters were known as jumper tenders, and they had to be both alert and strong because steering required them to lift and turn the sleds by hand as they bounced along the twisting, debris-filled track. After winter ice and snow smoothed out the roads, larger horse and sled teams would haul the collected bark from the yards to rail lines, lake shores, or the tannery. As for the stripped trunks, loggers might haul them out of the woods and cut them into lumber, but more likely they would be left to rot.[20]

As I've discussed, hemlocks grow slowly; once they are destroyed, it takes many decades to replace them. This meant that both the bark harvesters and the hemlock-fed tanneries were living on borrowed time. According to the *Tenth Annual Report of the Bureau of Industrial and Labor Statistics for the State of Maine* (1896), harvesting bark "puts a large amount of money in circulation, yet it is a transient business":

For, unlike spruce, when once cut over, hemlock does
not readily reproduce itself, and when the bark supply is
exhausted, . . . the site is abandoned and the buildings left to
decay. Such has been the fate of nearly all the large tanneries
in the older settled sections of the State . . . Immense tracts
of hemlock which have been stripped during the last forty
years with such a reckless hand can never be reproduced,
and the area yet remaining will, in a very few decades at
most, be left barren of this valuable growth.

The report predicted that "the [tanning] industry will disappear from our midst."[21]

Even before 1896, the Shaw brothers were confronting this
reality. The family had built five of the nation's biggest tanneries
within the confines of a small forested area in eastern Maine,
and by the early 1880s they were struggling to supply all of them
with ooze. To compensate, they built extract mills farther away
from the tanneries, for the sole purpose of extracting and supplying tannins for leather making. This change was a sign of
problems to come and suggested that the Shaws had expanded
their tanneries too rapidly. Still, at the time of William Shaw's
death in 1882, the family appeared to be in complete control of
an empire that included operations in Maine, New York, and
Canada. Collectively, the Shaw companies were probably the
biggest producers of sole leather in the world, with about twenty
tanneries in the United States and Canada generating 13,000
tons of leather per year. Then, in late July 1883, the Shaws made
a shocking announcement: they were suspending operations due
to "involuntary insolvency." The company did not have enough
cash to pay its creditors. It was broke.[22]

The failure of the country's largest leather-making firm made
headlines. The *Boston Daily Advertiser* reported that the news was
"so entirely unexpected, and . . . involve[d] amounts so gigantic,
as to cause a profound sensation . . . , the effects of which are
likely to be felt directly in Boston, throughout Massachusetts,

in northern New England and New York, and in Canada."[23] The immediate cause of the collapse was said to be the failure of C. W. Copeland and Company, one of the largest shoe-and-boot firms in Boston. The Shaws had made loans to Copeland; they held promissory notes for hundreds of thousands of dollars. So when Copeland failed and could not meet its obligation to the Shaws, then the Shaws, in turn, found it difficult to pay off those who had loaned money to them. Building the tannin extract mills had further deepened their financial hole.

But as the Shaws responded to their creditors, it became clear that their problems went beyond bad loans to Copeland.[24] Shortly before their company failed, the Shaws had claimed assets of between $5 million and $6 million, which had made them seem like a solid investment. But as creditors arrived to evaluate the company's goods, the family now asserted that their assets amounted to no more than $3 million. How had millions of dollars in assets evaporated? According to the company, "we have made large reductions in the estimated value of tanneries, [because] the tanneries are very large ones, and the bark resources are insufficient to run them to their full capacity; hence they are no more valuable than much smaller tanneries would be in the same locality." That is, the Shaws did not have enough hemlocks left to produce leather at peak rates, and the investment in extract mills had added debt while failing to change this equation. As a result, the tanneries were now worth far less than they once had been. It was clearly in the family's interest to devalue their holdings to reduce their obligation to creditors, and a newspaper editor in Boston speculated that "failure was a masterly scheme for the safe reduction of a business grown too large for its managers." But it was also true that the Shaws had engaged in tree mining—removing the hemlocks far faster than they could be replaced. They had depleted the forests to sustain the production of leather, and now much of what they owned had little value.[25]

Despite these financial troubles, the Grand Lake Stream

tannery was not quite ready for the knackers, and the hemlock forests were not yet completely gone. Creditors in Boston and Portland took possession of the property in hopes of getting at least some of their money back, and within a few years the operation was tanning hides again, though at a lower rate. In 1887, a major fire caused by sparks from the furnace destroyed the tanyard, the beam room, and the ten-story dry loft, which acted as a giant chimney as it burned. But, astonishingly, someone at an insurance company had written a policy for the highly flammable facility. So the tannery was rebuilt and continued to produce leather into the 1890s. However, its worth was depreciating fast as the surrounding hemlock forest disappeared. In 1898, thirty years after the facility had opened, the end arrived—a year or two after the International Leather Company purchased the works. Anything International Leather could not use in other locations was purchased by junk dealers, who hauled away boilers, engines, bark mills, hide mills, rollers, and piping. Thrifty locals pulled apart buildings, saving the materials for sheds and firewood. Another fire burned much of what was left, and soon all that remained of the world's biggest tannery were a few stone foundation walls, caved-in sweat houses, partly destroyed tanning vats filled with stagnant water, and the eighty-foot brick chimney from the furnace.[26]

As for the Shaws, some remained in Maine after the 1883 failure, managing tanneries in other parts of the state. Others, including Fayette Shaw, took their knowledge of the leather business to the hemlock forests of Wisconsin, where they operated a new set of tanneries. In 1906, after a fire destroyed an operation in Philips, Wisconsin, a reporter for the *Milwaukee Sentinel* asked Shaw if he would rebuild. He replied, "I am seventy years old. I have actively engaged in tanning leather for more than fifty years. This is the seventh tannery I have had purified by fire. By the Eternal, I am too old a man to be beaten in this way. The tannery will be rebuilt!" He was true to his word.[27]

But the twentieth century brought American industry new methods for tanning leather. In the mid-1800s, technologists in Europe had discovered that hides could be tanned with chromium (III), or Cr^{3+}. Like tannins, chromium could cross-link collagen fibers in the skins to produce leather, but it could also be shipped to wherever the hides were produced. Now there was no need to transport hides for thousands of miles to the source of the tanning chemicals, no need to cut down tanbark trees, and no need to build tanneries in the forests. During the 1900s, chromium tanning almost completely replaced tree-bark tanning in the leather industry. While this created its own set of significant pollution problems, it also saved the slow-growing, hard-to-replace eastern hemlock. Or, rather, it saved the trees until the hemlock woolly adelgid made its entrance onto the stage, but that's another story (see the appendix).

Appendix

Finding and Identifying Conifers at Acadia National Park

While the conifers of the New England–Acadian Forest grow throughout the ecoregion, Acadia National Park is an ideal location for observing them in a variety of habitats. As the park's founding father George Dorr said, it offers "a permanent exhibit of this forest growing under original conditions." Some of the species described in this appendix are not discussed in the preceding chapters, but all are important in the ecology of the region, and all grow on the coast of Maine.

The following is a guide to finding and identifying these conifers. If you are interested in a more expansive one, look for *Forest Trees of Maine*, published by the Maine Department of Agriculture, Conservation, and Forestry.

Key to Conifer Genera at Acadia National Park

Needle-shaped leaves—**Pine Family** (Pinaceae)

    One needle per cluster

        No sterigmata or bumps on twigs—**Fir** (*Abies*)

        Needles on sterigmata or bumps on twigs

            Needles spiral around twigs—**Spruce** (*Picea*)

            Needles in flat, horizontal sprays—**Hemlock** (*Tsuga*)

    More than one needle per cluster

        Needles of clusters of two, three, or five—**Pine** (*Pinus*)

        Needles in clusters of eight to ten—**Tamarack** or **Larch** (*Larix*)

Flattened, scaly leaves—**Cypress Family** (Cupressaceae)

    Individual leaves are tiny, flattened, scaly, and overlapping
    —**Cedar** (*Thuja*)

Spruce species (*Picea* species)

Identifying species within the spruce genus can be challenging. There are few clear differences among the three spruce species (red, white, and black) found at Acadia National Park, and the range of values for traits such as cone and needle length often overlap. Red and black spruce are so similar that many nineteenth-century botanists treated them as one species, known then as black spruce. In addition, red and black spruce trees can hybridize and produce trees with intermediate characteristics.

Thus, it's useful to remember that there are also differences in abundance, habitat, and location. The vast majority of spruce trees on Mount Desert Island are red spruces, and they are found all over the island; when in doubt, assume that you're looking at a red spruce. White spruce trees are abundant along the southeastern coast at Otter Point, the Ocean Path, Great Head, Schooner Head, and Wonderland as well in some of the interior red spruce forests. Black spruce is the least common type of spruce at Acadia. Unlike red spruce, they are adapted to damp and boggy soils, and most are found in the southwestern corner of the island.

FIGURE 20. Comparison of spruce species cones: black spruce (left), red spruce (middle), white spruce (right).

TABLE 1. COMPARISON OF THE TRAITS OF THREE SPRUCE SPECIES

	RED SPRUCE	BLACK SPRUCE	WHITE SPRUCE
Needle Length	0.25 to 0.70 inches (0.6 to 1.8 cm)	0.25 to 0.50 inches (0.6 to 1.3 cm)	0.50 to 0.75 inches (1.3 to 1.9 cm)
Needle Description	Shiny, yellow-ish-green Stiff, slightly curved, narrower and sharper at tip than black spruce	Dull green Flexible, straighter, wider and blunter at tip than red spruce	Dull green Wider and blunter at tip than red spruce, stiff with odor of cat urine when crushed
Twigs	Covered with tiny hairs	Covered with tiny hairs	Hairless
Bark	Reddish-brown scales	Grayish-brown scales	Light gray-brown scales
Cone Length	1.2 to 2.0 inches (3.0 to 5.1 cm)	0.5 to 1.4 inches (1.3 to 3.6 cm)	1.3 to 2.2 inches (3.3 to 5.6 cm)
Cone Description	Rich chocolate brown; scales stiff; cone wider or more rounded in the middle than white spruce cones	Dark brown; scales stiff and brittle; cone egg-shaped to nearly spherical when scales open	Light brown to tan; scales thin and flexible so that open cone easily compressed; closed cone cylindrical (much longer than wide)

Red Spruce (*Picea rubens*)

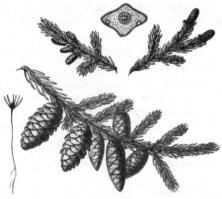

FIGURE 21. Red spruce, including twigs, foliage, cones, a seedling, and a cross-section of a needle.[1]

Identification

Red spruce leaves grow as separate individual needles attached to twigs by bumpy, peglike sterigmata and arranged in sprays that spiral around the twigs. They are a shiny yellowish green and are 0.25 to 0.70 inches long. There is much overlap in the range of needle lengths among the spruce species, but, on average, red spruce needles are longer than black spruce needles and shorter than white spruce needles. Red spruce needles are also more curved, and they are narrower and sharper at the tip than black spruce needles are.

Red spruce twigs are covered in tiny hairs, about 0.2 to 0.3 millimeter in length. It can be very difficult to see them without a 10X magnification jeweler's loupe (available from many sources; usually costing less than $30). Black spruce twigs also have tiny hairs, but white spruce twigs are hairless, so the presence of twig hairs is a good way to distinguish red spruces from white spruces.

The cones have stiff scales (white spruce cones have flexible scales), are a rich chocolate brown in color, are 1.2 to 2.0 inches long, and are wider in the middle than on the ends, especially when still closed. There is overlap in the range of red spruce and black spruce cone lengths, but black spruce cones are rarely longer than 1.5 inches. The bark is a dark reddish-brown and broken into potato

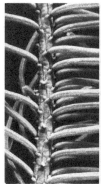

FIGURE 22. Left: Spruce twig showing sterigmata. Center: Red spruce twig magnified to show tiny hairs (gray arrows). Right: White spruce twig magnified to show the absence of tiny hairs.

chip–sized scales or plates; white and black spruce barks have a more grayish tint. Older, taller red spruce trees can be distinguished from the hemlocks by their bark: spruce tree bark is broken into scales or plates, while hemlock bark forms long ridges.

Distribution at Acadia National Park

Red spruce is a key diagnostic species for the Acadian Forest ecoregion, and it's impossible to miss this species at Acadia National Park as it accounts for about 40 percent of the park's trees. In the southern and western parts of the park, red spruce is often the dominant tree species, and numerous trails and carriage roads pass through extensive stands of older red spruce trees. For example, some of the tall red spruce trees in the mature forest along the eastern section of the Asticou and Jordan Pond Path have trunks that are up to two feet in diameter (see map, Figure 28). Given the rate of spruce tree growth in coastal Maine, these large trees could be 150 years old or more.

In contrast, most of the red spruce trees in the mixed hardwood-conifer forests growing along the Witch Hole Pond carriage roads are relatively younger and shorter. A vegetation map from the 1920s shows that this area was once covered by "mixed conifers," and it's likely that red spruce was a dominant or more common species at that time.[2] However, most of these trees were killed by the massive fire that swept through the northern part of the park in 1947. The fire created conditions favoring species that sprouted from stumps, produced seeds that dispersed over long distances, or grew well in drier soils and under higher light conditions. As a result, the early twentieth century's mixed conifer forest was replaced by a blend of aspens, birches, maples, oaks, and white pines along with some red spruce. Red spruce is long-lived and very shade tolerant, and perhaps the smaller spruce trees now found in the Witch Hole Pond area will grow up to return this species to dominance. The hope is that climate change will not derail this successional process by creating conditions unfavorable to red spruce, but current climate models suggest that this very type of change is likely to occur at least by end of the twenty-first century.

White Spruce (*Picea glauca*)

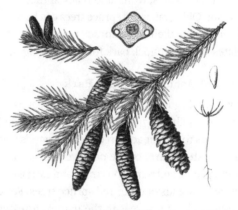

FIGURE 23. White spruce, including twigs, foliage, cones, a seed, a seedling, and a cross-section of a needle.

Identification

White spruce leaves grow as separate individual needles and are attached to twigs by sterigmata. Needles are a dull dark green and 0.50 to 0.75 inches long—on average, a little longer than red and black spruce needles. Needles on the lower side of white spruce twigs curve strongly upward so that they often appear to be all on the upper side. The surface of the twigs is smooth and hairless, a feature that can be confirmed using a 10X magnification jeweler's loupe. Absence of tiny hairs is an effective way to distinguish white spruce from the hairy-twigged black and red spruce. Needles may also have a sharp aroma when crushed, similar to cat urine, although odor strength can vary with foliage age.

The cones have flexible scales (red and black spruce cones have stiff scales), are tan or light brown in color, are 1.3 to 2.2 inches long, and are cylindrical when cones are closed. Open cones are more rounded and can resemble open red spruce cones, but they are longer and more cylindrical than red spruce cones, and the flexible scales make it easy to compress or squash an open white spruce cone between thumb and index finger.

The bark is grayish-brown and broken into potato chip–sized scales or plates; red spruce bark is more reddish. White spruce trees are usually more conical or Christmas tree shaped than black spruce, although, when growing on headlands, they may have a wind-shaped, irregular, steeple-like appearance. Bark with scales and plates distinguishes white spruce from hemlocks, which have bark with long ridges.

Distribution at Acadia National Park

White spruce is a boreal, or northern forest, species more at home in Canada than on the coast of Maine. On Mount Desert Island, the species can be found near the southern edge of its range, and these trees are much less abundant in Acadia National Park than red spruce is. White spruce grows in the interior of the island, often mixed into red spruce forests, but the species accounts for a larger portion of the conifers in habitats along the island's southwestern and southeastern edges, where the trees grow in shore-hugging bands. White spruce stands can be found along the rocky headlands sections of the Ship Harbor and Wonderland trails (see map, Figure 26). They also grow in an almost unbroken band on the coast from Great Head south to Sand Beach and then along the Ocean Path past Thunder Hole to Otter Cliffs and Otter Point, where some of the trees stand on the headlands like coniferous sentinels. The devastating 1947 fire reached Great Head but not the Otter Point peninsula, so in this part of the park the oldest and tallest white spruce are growing along Otter Point at the southern end of the Ocean Path.

A short distance inland from the shoreline band, the dominant spruce species usually changes over to the ubiquitous red spruce. The botanists Duncan Johnson and Alexander Skutch observed this pattern—a shoreline band of white spruce bordered by inland red spruce—at Otter Point in the 1920s.[3] Their observations provide evidence that this spruce species distribution has persisted for much longer than a century. The pattern may be the result of a cooler, damper microhabitat at the ocean's edge, which could favor the boreal forest white spruce. Or white spruce may be better than red spruce at colonizing the seaside gaps created by storms. Because white spruce is less shade tolerant and has a shorter lifespan than red spruce does, its continued presence at Otter Point may depend on these disturbances. George Dorr noted that the white spruce "is the only evergreen tree on the coast . . . whose roots will maintain their hold, growing vigorously, on a surf-cut bank till it is washed away," a trait that could give it an advantage over other conifers in headlands habitats.[4]

Black Spruce (*Picea mariana*)

FIGURE 24. Black spruce, including twigs, foliage, cones, a seedling, and a cross-section of a needle.

Identification

Black spruce leaves grow as separate individual needles and are attached to twigs by sterigmata. The needles are a dull dark green and are 0.25 to 0.50 inches long—on average, a little shorter than the needles of red and white spruce, although there is overlap in needle length among the species. Black spruce needles are also straighter and wider, and they are blunter at the tips than red spruce needles are . Like red spruce, the black spruce twigs are covered in tiny hairs that can be seen with a 10X magnification jeweler's loupe; the presence of twig hairs can be used to tell black and white spruces apart.

The cones have stiff and brittle scales (white spruce cones have flexible scales), are dark brown in color, are 0.5 to 1.4 inches long, and are egg-shaped when closed and nearly spherical when dry and open. Compared to red and white spruce, black spruce cones have the shortest average length and rarely exceed 1.5 inches in length.

The bark is grayish-brown and broken into potato chip–sized scales or plates; red spruce bark is more reddish. When growing in the open, older black spruce trees usually have a narrow, cylindrical, steeple-like shape, with short branches and an open irregular crown. Some may carry tight clusters of old, gray, weathered cones near the crown, visible with binoculars. Red spruce trees growing in the open are more conical and look more like classic Christmas trees.

The lower branches of black spruce are usually strongly drooping. If they touch the ground, a new tree may be produced by a process called layering. Not all black spruce trees reproduce by layering; but given that this habit is rarely seen in red spruce, it is a good indicator that a tree is a black spruce. This method for creating a new tree begins when a downward-growing, ground-touching branch of parent black spruce becomes covered in leaf litter and sphagnum moss. Where the branch touches the ground, new adventitious roots grow out and down. These provide a new source of nutrients for the branch that is separate from the parent tree's roots. The branch now begins to act like a small, semi-independent sapling, gradually changing its direction of growth from roughly horizontal like the other lower branches to vertical or upward pointing. Eventually, the new tree can survive using its adventitious root system alone, and over time the physical connection and the flow of nutrients from the parent tree ends. If the older tree dies, the tree produced by layering can continue to live and grow. Black spruce trees that reproduce by layering will have lower branches diving into the ground and smaller trees popping out of the ground a few feet from the base of the trunk along the line of the parent tree's buried branch. Older trees that are a product of layering can be identified by a gradual or swooping curve in the base of their trunks, a sign that they began by growing from a buried curved branch. Occasionally, red spruce trunks will also have a curve at the base of the trunk if the tree began life on or near a boulder, but the obstruction can usually be seen and the bend at the base is usually close to a ninety-degree angle as opposed to the gradual swooping bend rising out of a rock-free surface, as seen with a layering black spruce.

FIGURE 25. Curved trunks on black spruce trees produced by layering.

Distribution at Acadia National Park

Black spruce is also known as bog spruce or swamp spruce, names that reflect its adaptation to wetlands soils. It's abundant in Canada and is the official tree of the province of Newfoundland and Labrador. However, the species is not common on Mount Desert Island, where it is near the southern limit of its range in the eastern United States, so finding black spruce trees can take a little effort. They do grow in a few sphagnum bogs, on the marshy margins of some lakes, and on a few of the headlands where depressions in the rocks form poorly drained peaty pockets. At Acadia National Park, most black spruce trees are found in the southwestern part, in the Seawall area, and the Ship Harbor Trail is a good place to look for them (see map, Figure 26).

The Ship Harbor Trail begins at a parking area along Route 102A. From the trailhead, take the path to the first trail junction, about 0.1 miles from the parking lot. Make a right (toward the west) turn to follow the northwestern section of the trail's figure-eight loop. A short distance past the turn, the trail enters a forest in which relatively abundant black spruce trees are mixed in with white and red spruces. Most of the black and white spruces are on the right or ocean side of the trail, while most of the red spruces are on the left or inland side of the trail. This path is one of the few places in the park where you can see all three spruce species in the same small area.

The black spruce trees living in the less crowded spaces along the rocky edge of the Ship Harbor can be recognized by their narrow, cylindrical, steeple-like, or irregular shapes. Some may carry tight clusters of old, gray, weathered cones near the crown; unlike red spruce trees, black spruce trees may retain their cones for several years. Within the denser forest to the left, it can be difficult to see the overall shape of the trees, and the crowded red spruce may have a narrower growth form. However, the presence of black spruce may be revealed by the tree's shorter, more rounded cones. Because the black and red spruce trees are growing close together, the cones are often intermingled on the ground; but if a given tree is a black spruce, most of the cones and needles under the tree should be from this species. When looking at cones, note that their size can vary and that black spruce and red spruce can hybridize, producing trees with cones whose measurements fall between the two parent species.

As I've mentioned, black spruce can also be identified by layering, and there are many examples of this along the Ship Harbor Trail—for instance, on both sides of the path at a point where a small rocky outcrop juts out into Ship Harbor, about halfway through the northwestern section of the trail's loops. On the right (western) side of the trail are a few black spruce trees with multiple lower branches, growing on the rocky ground or clinging to the side of the outcrop. On the left or eastern side, there's a large tree growing on better soil that appears to have produced four or five good-sized offspring trees by layering.

Beyond the junction in the middle of the figure-eight loop, there appear to be fewer black spruce trees, although there are still some on the ocean side, most obviously on the rocky outcrops around the point where the path emerges onto Ship Harbor headlands. Continuing south, the headlands are dominated by white spruce. Then after turning inland at the southern end of the trail, the path is most notable for the numerous tamaracks in its southeastern section and for a maturing red spruce–dominated forest along its northeastern section. The Ship Harbor Trail is some distance from the more popular eastern side of Mount Desert Island, but it also has an unusual mix of conifers and is an ideal place to see uncommon species such as black spruce and tamarack. (And it has some nice tidepools at its southern end.)

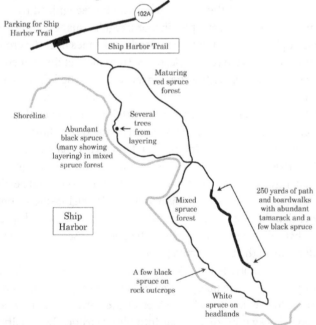

FIGURE 26. Ship Harbor Trail. Visitors can see three spruce species (black, white, and red) and three boreal forest species (black and white spruce and tamarack).

Pine species (*Pinus* species)

There are four pine species at Acadia National Park, although the red pines are now rare due the destructive activity of the red pine scale. These species can be distinguished from each other on the basis of needle length and the number of needles per cluster.

Eastern White Pine (*Pinus strobus*)

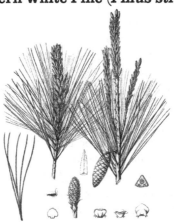

FIGURE 27. White pine, including twigs, foliage, cones, a seedling, and a cross-section of a needle.

Identification

White pines are easy to identify because they are the only Acadia pine species with clusters of five needles, 3 to 5 inches in length. Mature cones are slender and about 4 to 8 inches long; other Acadia pines species have much shorter cones. When trees are young, the bark is smooth and gray-green; in older trees, the bark is brown and divided into broad ridges.

Distribution at Acadia National Park

White pines are abundant at Acadia National Park. For example, they are easy to find along the carriage roads near Witch Hole and Breakneck Ponds in the northern part of the park, where most of the forest burned in 1947. This huge fire consumed about 15,000 acres, including about half of the national park land located east of Somes Sound. The presence of white pines along these carriage roads illustrates the tree's role as a pioneer species in habitats opened by fire, logging, or farming. After the big burn, aspens, birches, red maples, red oaks quickly colonized the area, arriving as wind-born seeds or regrowing from stumps and root crowns to produce a mixed hardwood forest. But white pines are also good at pioneering in open habitats, and the seed-generated pine seedlings and saplings are better adapted to post-fire environments than are young red spruce plants. As a result, white pines are among the oldest and tallest conifers growing among the hardwoods in this northern part of the park.

White pines of many ages are also a component of the older red spruce forests in the southern sections of the park, which were untouched by the 1947 fire. For example, they are found in the red spruce forest along the eastern segment of the Asticou and Jordan Pond Path. White pine seedlings and saplings struggle in the deep shade of mature conifer forests because the young plants need a minimum of 20 percent sunlight to survive. However, if wind or another disturbing force creates a light-filled gap in the conifer canopy, the pine saplings can shoot up at a faster rate than fir, cedar, and spruce saplings. This can be seen in the sunnier areas of the windthrow gaps, where the young white pines are outgrowing the red spruce saplings. This more rapid growth is essential to the pines' survival as they will lose out to the shade-tolerant conifers if they don't remain in the sun. Eventually, the white pines reach a final height that is greater than that of the other conifers, permanently ending the risk of being shaded. This adaptation—fast growth to towering heights in spruce forest gaps—is what produced the white pine islands in the seas of red spruce that were an early target of tree hunters (see chapter 4). Samples of this can be found about 0.7 miles from the trailhead of the Jordan Stream carriage road, where there are a few very tall, massive pines with trunks of up to three feet in diameter scattered among the red spruces (see map, Figure 28). These larger trees are likely to be well over 150 years old.

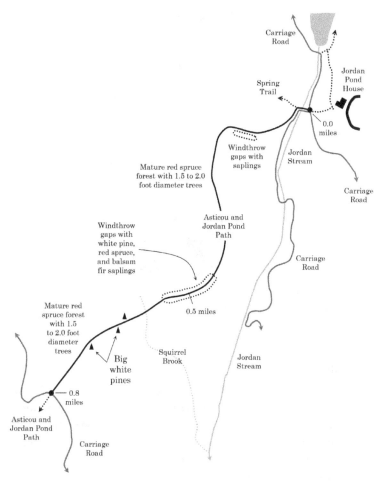

FIGURE 28. The Asticou and Jordan Pond Path. Visitors can see a mature red spruce forest, islands of massive pines, and conifer competition in windthrow gaps.

Pitch pine (*Pinus rigida*)

FIGURE 29. Pitch pine, including twigs, foliage, cones, seeds, a seedling, and a cross-section of a needle.

Identification

Pitch pine is the only Acadia pine species with leaves in clusters of three needles, 2.5 to 5 inches long and slightly twisted. Cones are prickly and 2 to 3.5 inches long. The trees have rough bark, which, on older trees, is divided into irregular dark gray plates and ridges. Trunks and large branches may have large epicormic buds with projecting clusters of needles.

Distribution at Acadia National Park

While the pitch pine range stretches from Maine to Georgia, different locations support different varieties of the species. Unlike most other pitch pine populations, the unique Acadian variety characterized by somewhat smaller trees that produce cones at an earlier age. These cones also have smaller seeds and more seeds per cone than most other pitch pine varieties.[5] Collectively, these reproductive adaptations allow the Acadian variety to disperse more rapidly over longer distances. The traits might have been of great value when the glaciers retreated from the coast of Maine about 15,000 years ago, creating opportunities for plant species to expand their ranges from south to north. There is genetic evidence that ancestors of today's Acadian pitch pine variety were growing on the continental shelf to the south on land that was exposed when huge amounts of the planet's water were tied up in ice sheets. The uncovered land functioned as a refuge for trees at a time when the Maine coast was buried in ice. As the ice melted, the pine populations used traits such as earlier reproduction and smaller winged seeds to expand their ranges northward as their Ice Age refuge was slowly drowned by the sea. Today, descendants of the continental shelf pines are alive and well, perched above the sea on rocky mountain tops.

At Acadia National Park, most pitch pines grow in harsh habitats, including rocky ledges, the upper sections of ridges, and mountain summits, where the trees grow in the nutrient-poor sandy or graveled soils that have formed in depressions in the granite bedrock. Groves of pitch pine may include a mix of other tree species, but it's not unusual to find near-pure stands of these pine along the broad crests of many of Acadia's mountains. The twin peaks of Kebo Mountain are dominated by pitch pine, and the southernmost peak is covered by thick stands of twenty- to thirty-foot-tall pines; these are some of the tallest pitch pines on the island. Additional pine-colonized summits include Huguenot Head, Champlain, Gorham, and Dorr mountains, among many others.

Jack Pine (*Pinus banksiana*)

FIGURE 30. Jack pine, including twigs, foliage, cones, seeds, a seedling, and a cross-section of a needle.

Identification

Jack pine needles are stiff, slightly twisted, and 0.75 to 1.5 inches long, and they grow in clusters of two needles. (Red pine needles are also in pairs but are much longer.) Cones are 1 to 2.5 inches in length, lack sharp spines and prickles, and may remain on trees for many years. The bark is gray to brown; on older trees, it has irregular rounded ridges or scaly plates.

Distribution at Acadia National Park

Jack pine is a boreal tree species that can survive in habitats with a frost-free period as short as fifty days; this is the shortest frost-free period requirement of any of Acadia National Park's conifers. The species is a significant component of the Acadian Forests of southeastern Canada but is not common at the park, where it is near the southern edge of its range in eastern North America and is found in isolated pockets. The park's fragmented jack pine populations appear on the Schoodic Peninsula and in a small patch of about eight acres growing along the spine of the

southern ridge of Cadillac Mountain. This latter population can
be reached via the park's Cadillac South Ridge Trail. For hikers
approaching from the southern end of that trail, jack pines first
appear at a point about 0.5 miles north of the northern junction
with the Eagle's Crag Loop Trail. (The half-mile between the
trail junction and the edge of the jack pine stand is dominated by
pitch pine.) Then, for the next 0.25 miles, the trail passes through
an area with both isolated jack pines and pines gathered in small
groves, until the jack pine stand ends about 0.25 miles south of
the junction of the Cadillac South Ridge Trail and the Canon
Brook Trail or about 1.5 miles south of the parking area on the
top of Cadillac Mountain.

The jack pines growing along the Cadillac South Ridge Trail
are mixed with or adjacent to patches of pitch pines. Here, the
two types exist in a tension zone—that is, in an area in which
two or more species reach their range limits from opposite directions. It's the southern limit for jack pine and the northern limit
for pitch pine. In such locations, one type of ecological community blends into another, and species from both groups are
found in the same location. Like jack pine, pitch pine does well
in habitats with little shade, poor soils, and frequent fire. And
when two intermingling species are adapted to similar habitats,
they almost inevitably compete with each other.

Who comes out ahead on Mount Desert Island? Usually, pitch
pine. It's much more common and more widely distributed on
the island than jack pine is. But jack pines do grow here, and
research done in the 1990s by Michael Greenwood and others
showed that the two types have subtle differences in adaptation. These distinctions seem to enable both species to endure
in slightly different microhabitats on Cadillac Mountain.[6] In
seed beds designed to mimic the fast-drying granite-sands soils
of Cadillac Mountain, Greenwood found that jack pine seeds
germinated faster and at a high frequency than pitch pine seeds
did. Jack pines also allocated more biomass to roots than to
shoots or stems when compared with pitch pine. More root to

shoot translates into more below-ground biomass for absorbing water and less above-ground biomass in need of water. While this is ideal for tolerating drought conditions, it also means that the young trees are less able to acquire carbon and to make the plant biomass needed to compete against other species under good growing conditions.

These small variations between jack and pitch pines enable the two species to take different paths to success. Jack pines follow a germination and root-to-shoot strategy that improves survival and competitiveness in habitats with shallow, graveled, well-drained, drought-sensitive soils. On Cadillac Mountain, these pines are found most often in crevices in the granite and in locations with thinner soils. In contrast, pitch pine's greater investment in photosynthetic shoot tissue can be a winning approach at sites with slightly better soils where there may be more competition from other tree species. Cadillac's pitch pines tend to be found growing in circular hollows in rocks that contain deeper, moister soils. Thus, each competing species is adapted to slightly different microhabitats, and jack pines coexist with pitch pines by being just a little bit better in cracks, crevices, and other unpromising patches. Perhaps Cadillac's jack pines would relate to Thoreau's sentiment that he would "rather sit on a pumpkin and have it all to myself than be crowded on a velvet cushion."[7]

Red Pine (*Pinus resinosa*)

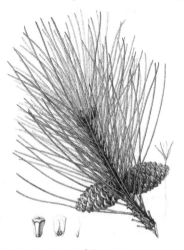

FIGURE 31. Red pine, including twigs, foliage, cones, cone scales, and seeds.

Identification

Red pine needles are 5 to 6 inches long and grow in clusters of two needles. Unlike jack pine needles, red pine needles snap or break cleanly when bent in half. Cones are egg-shaped, are 1.5 to 2.5 inches in length, and lack prickles. The bark is composed of reddish-brown plates or broad scaly ridges.

Distribution at Acadia National Park

Red pines are more abundant around the Great Lakes and the Saint Lawrence River than in New England, where they often grow in small stands on rocky ridges and along lakes. On Mount Desert Island, this species was always less common than many other conifer species, despite being well adapted to some of soil types found on the island, including acidic, low-nutrient, well-drained, sandy soils of glacial origin. Unfortunately, a tiny insect called the red pine scale has now killed nearly all of the island's red pine trees. At almost every location where they once grew in Acadia National Park, all that remains are bare branches and trunks with reddish-brown bark.

Unless a particular stand of dead trees creates a significant hazard to visitors, the Park Service prefers to leave them in place until they decay away. For now, dead and dying stands can be found along the carriage roads between Eagle Lake and Witch Hole Pond. To take a sad tour of the dead red pines, begin at the parking area for the Eagle Lake carriage roads on Route 233, and take the carriage road that leads to the north toward Breakneck and Witch Hole ponds. About 0.65 miles north of the parking lot, there's a stand of dead red pines on the lefthand (western) side of the road, where the land juts out a bit into one of the Breakneck ponds. Then over the next 0.25 miles, dead pines can be seen almost continuously on the righthand (eastern) side of the road, mixed in with a variety of living species. At about 0.9 miles from the parking area, the carriage road turns to the right (east); and as the road approaches the signpost 4 intersection, dead pines appear again on the righthand (southern) side of the road. A left (north) turn at signpost 4 leads past another stand of dead pine directly across the road from Halfmoon Pond (on the right, or eastern, side). For even more dead red pines, continue another 0.8 miles to the marshy southern end of Witch Hole Pond, where there are more on the left (western) side of the road mixed in with other tree types.

Other conifer species
(*Abies, Tsuga, Larix,* and *Thuja* species)

Each of the remaining conifer species at Acadia National Park is in its own genus.

Balsam fir (*Abies balsamea*)

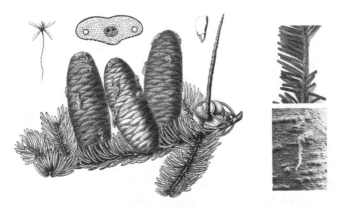

FIGURE 32. Balsam fir. Twigs and cones (left), twigs without sterigmata (upper right), resin blisters with dried resin (lower right).

Identification

Balsam fir needles grow as individual needles (not in clusters), and are 0.5 to 1.5 inches long, dark green in color, and whitened on the undersides. Trees may appear to be spruce-like, but fir needles are attached directly to the twigs, not by peg-like sterigmata. Thus, after the needles have been removed, the twigs are smooth. The absence of sterigmata distinguishes balsam fir from hemlock and spruce species. Balsam fir cones are barrel-shaped, 2 to 4 inches long, and erect or upright, meaning that they project above the branches. The bark of young trees is pale gray and smooth with resinous blisters that can be cut open to open to collect Canada balsam (see chapter 6). Older trees have rougher bark and usually lack blisters.

Distribution at Acadia

Balsam fir is the only fir species native to northeastern North America, and it is widely distributed in eastern Canada and the northeastern United States. With its preference for colder climates, balsam fir was probably among the first conifer species to colonize Mount Desert Island after the glaciers retreated about 15,000 years ago. Today, at Acadia National Park, fir trees are widely distributed from sea level at Otter Point to the top of Cadillac Mountain.

The species is strongly associated with red spruce, both at the national park and in numerous other locations in the New England–Acadian Forest ecoregion. Beyond the national park, young fir trees often outnumber young spruce trees in areas cleared by logging, a result that can frustrate foresters because they consider red spruce to be more commercially valuable. On Mount Desert Island, the very shade-tolerant firs usually occur as understory saplings or as small trees within the mature red spruce forests. These spruce-fir forests are widespread throughout the southern and western sections of the national park, and you can find them along Jordan Stream, the Asticou and Jordan Pond Path, the carriage roads near Upper Hadlock Pond and Cedar Swamp Mountain, and in many other locations as well. Balsam firs are also found mixed in with white spruce. At the Otter Point (southern) end of the Ocean Path, the firs compete with white spruce along a rocky shoreline where strong winds created gaps in the spruce forest, which the firs have exploited. Here, most saplings and smaller trees are balsam firs while the taller, older trees are usually white spruces.

Eastern Hemlock (*Tsuga canadensis*)

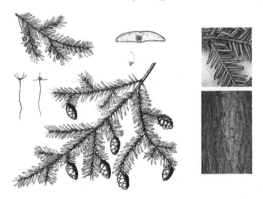

FIGURE 33. Eastern hemlock, including twigs, foliage, cones, seeds, a seedling, and a cross-section of a needle.

Identification

Hemlock needles are short, 0.4 to 0.8 inches long, with dark-green upper surfaces and whitened undersides. They grow as individual needles on sterigmata in flattened horizontal sprays. The cones are small, 0.75 to 1.5 inches long, and egg-shaped. The bark is scaly and gray-brown when the trees are young; older trees have dark gray to brown bark with long ridges and deep furrows. When trees are too tall for you to see the needles clearly, you can distinguish hemlock from spruce species by the bark: hemlock bark has long ridges, while spruce bark is broken into plates or scales.

Distribution at Acadia National Park

Hemlocks prefer cool, moist soils, and in Acadia they are most likely to be found in older forests along the banks of streams, ponds, and lakes and in the damp soils at the bases of ridges. For example, these trees are abundant in the mature red spruce forests that grow along Jordan Stream. You can explore these forests using either the carriage road or the hiking path that parallels Jordan Stream.

In the northern part of the park, the aptly named Hemlock Road is a good place to find hemlocks, including several impressive trees that are more than two feet in diameter. To find them, park at the Sieur de Monts parking area and follow the sidewalk along the fence in front of the Wild Gardens of Acadia at the northern (righthand) end of the gardens. A short distance beyond the northern end, turn left onto a wide gravel path that follows the northern border of the gardens. Take this path to junction of the Hemlock Road and the Jesup Path (the boardwalk path), and then follow the Hemlock Road straight ahead.

Along the Hemlock Road, the woods to the left (west) at the base of Dorr Mountain are filled with hemlocks of assorted sizes mixed in with several hardwood species such as yellow birch, which is often associated with hemlocks. About a quarter mile from the Jesup Path, near the junction with the Stratheden Path on the left (marked with a large carved boulder), there are several massive hemlock trees with trunks of up to two and a half feet in diameter. More very large hemlocks are located along the first (southern) part of the Stratheden Path. These trees are likely to be at least two hundred years old and must have survived the devasting 1947 fire. How they did so is a mystery as the thin-barked, shallow-rooted hemlocks are not generally very resistant to fire. It's possible that the steep slopes of Dorr Mountain to the west provided some shelter from the flames, which were driven by high winds blasting across the island from the west and northwest.

Today, the greatest threat to the park's hemlocks comes from the hemlock woolly adelgid, a tiny, aphid-like, sap-sucking insect.[8] Native to Japan, it was probably accidentally imported on live plant material from an Asian hemlock species. The adelgid was first documented in the eastern United States on a museum specimen collected in Virginia in 1951. Since then, the pest has killed huge numbers of hemlocks throughout much of the southern part of the tree's range. Adelgids have now been found at Acadia National Park, and the Park Service is developing plans to cope with the invasive species.

Tamarack or Eastern Larch (*Larix laricina*)

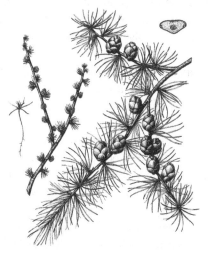

FIGURE 34. Tamarack or eastern larch, including twigs, foliage, cones, a seedling, and a cross-section of a needle.

Identification

Tamarack needles grow in clusters of eight to ten soft needles on spurs, 0.75 to 1.5 inches long, which are bright green during the growing season and yellow in autumn. Then the needles fall from this deciduous species. Cones are small, 0.5 to 0.75 in length, and broadly egg-shaped to spherical; unlike the needles, they usually remain attached throughout the winter. The bark of young trees is thin, smooth, and gray; on older trees, it is scaly and reddish-brown.

Distribution at Acadia National Park

Tamarack is a classic boreal forest species, along with jack pine, white spruce, and black spruce. The trees can tolerate a wide range of soil conditions, but they are most strongly associated with moist to wet, acidic, organic soils such as those found in peat-rich sphagnum bogs and in forested swamps, where woody material in the soil is slow to decay. On Mount Desert Island, the species is not common on the eastern side of the island, although a few trees may be found around the marshy margins of some ponds and lakes. Tamaracks are more abundant in the southwestern part, where they grow on damp or boggy ground, often accompanied by the wetlands-loving black spruce, the tree's most frequent conifer companion.

Tamaracks appear along the Ship Harbor Trail, one of the few locations in the park where it's easy to find three major conifer species of northern Canada's boreal forest: tamarack, black spruce, and white spruce (see map, Figure 26). The tamaracks are distributed along 250 yards of a path with three boardwalks in the eastern half of the southern loop of this figure-eight trail. Here, the tamarack's neon-green foliage stands out against the darker greens of the other conifers and the hardwoods; in the autumn, its deciduous needles turn bright golden yellow. Oddly, instead of being found in a near sea-level marsh, this tamarack patch is located on low broad hill—only about thirty feet above sea level but still higher in elevation than any of the land immediately around it. One would expect water to drain off its top, but perhaps there is a depression in the underlying granite bedrock that traps and holds rainwater and snowmelt in a rock bowl so that the soil on this plateau remains damp enough to favor tamaracks.

Northern White Cedar (*Thuja occidentalis*)

FIGURE 35. Northern white cedar, including twigs, foliage, cones, a seedling, a magnified leaf tip, and a cross-section of a needle.

Identification

Individual leaves are tiny, about 0.10 to 0.15 inches long, and they are flattened, scale-like, overlapping, and completely cover the twigs. Cones are egg-shaped, about 0.5 inches in length, with only a few scales per cone. The bark of younger trees is smooth; but as they mature, it become fissured with grooves separating fibrous, shredding ridges.

Distribution at Acadia National Park

Northern white cedars are widespread at Acadia National Park, where they are found in typical white cedar habitats such as the banks of streams, ponds, and lakes; in the moist soil of upland forests; and in wetlands known as cedar swamps. Trails and carriage roads that pass through cedar-favoring environments include the Jordan Pond Path, the Asticou and Jordan Pond Path, the carriage road that parallels Jordan Stream south of Jordan Pond, and the carriage roads along Bubble and Upper Hadlock Ponds, among many other locations.

Though associated with damp soils and wetlands, cedars can also be found growing on cliffs, rock outcrops, and summits with thin, dry, coarse, sandy, or graveled soils. At Acadia, it's not unusual to see them growing out of cracks in exposed granite or on mountaintops, including Cadillac Mountain. At Bubble Pond, they not only grow in the damp soil on the pond's western shore but can also be seen clinging to the steep, rocky western face of Cadillac across the pond from the carriage road. At first glance, such harsh habitats would not appear to be good hunting grounds for those stalking the damp soil–associated *Thuja occidentalis*. But these trees can survive at sites at both ends of the soil-moisture spectrum. They compete well in damp habitats where other species may be restricted by a high moisture content. Although they are less competitive and may be eliminated at better sites, they come into their own again on dry, stony slopes. By doing relatively better in challenging environments that inhibit the growth of other tree species, white cedars may become a dominant part of the plant communities in swamps or on rocky ledges and peaks, especially those that have seeps or water-collecting depressions in the bedrock.

Notes

Introduction: The New England–Acadian Forest Ecoregion

1. Henry David Thoreau, *The Maine Woods* (Boston: Ticknor and Fields, 1864), 87.
2. Josh Noseworthy and Thomas M. Beckley, "Borealization of the New England–Acadian Forest: A Review of the Evidence," *Environmental Reviews* 28 (2020): 284–93.
3. George Dorr, *The Acadian Forest* (Bar Harbor, ME: Wild Gardens of Acadia, 1917), 3–4.

Chapter 1: A Tree of Life for French Explorers

1. Jacques Cartier, *The Voyages of Jacques Cartier*, ed. Ramsey Clark (Toronto: University of Toronto Press, 1993), 76–77.
2. Cartier, *Voyages*, 78.
3. Cartier, *Voyages*, 79.
4. Cartier, *Voyages*, 80.
5. Cartier, *Voyages*, 80.
6. Laurier Turgeon, "French Fishers, Fur Traders, and Amerindians during the Sixteenth Century: History and Archeology," *William and Mary Quarterly* 55 (1998): 585–610; Bruce Trigger, *Native and Newcomers: Canada's "Heroic Age" Reconsidered* (Montreal: McGill–Queen's University Press, 1985).
7. Cartier, *Voyages*, 109, 111.
8. Samuel de Champlain, *The Works of Samuel de Champlain*, vol. 1, ed. H. P. Biggar (Toronto: Champlain Society, 1922), 280.
9. Champlain, *Works*, 282–83.
10. Champlain, *Works*, 301.
11. Champlain, *Works*, 303–6.
12. Mary Moore, "Eastern White Pine and Eastern White Cedar," *Forestry Chronicle* 54 (1978): 222–23.
13. Marc Lescarbot, *The History of New France*, vol. 2, trans. and ed. W. L. Grant (Toronto: Champlain Society, 1911), 258.
14. Lescarbot, *History*, 344.

15. Lescarbot, *History*, 320, 270.
16. Champlain, *Works*, 448.
17. Lescarbot, *History*, 343.
18. Champlain, *Works*, 448.
19. Lescarbot, *History*, 343.
20. Cartier, *Voyages*, 100.
21. Don Durzan, "Arginine, Scurvy and Cartier's 'Tree of Life,'" *Journal of Ethnobiology and Ethnomedicine* 5 (2009): 5–20.
22. Jacques Mathieu, *L'annedda: L'arbre de vie* (Lille, France: Septentrion, 2009).

Chapter 2: White Pines and the King's Broad Arrow

1. John Springer, *Forest Life and Forest Trees* (New York: Harper and Brothers, 1851), 37.
2. Andrew Vietze, *White Pine: American History and the Tree That Made a Nation* (Guilford, CT: Globe Pequot, 2017).
3. Robert Albion, *Forests and Sea Power* (Cambridge: Harvard University Press, 1926).
4. Joseph Malone, *Pine Trees and Politics* (Seattle: University of Washington Press, 1964).
5. Robert Armstrong, letter to Charles Burniston, in *Calendar of Treasury Papers* (London: Eyre and Spottiswoode, 1889), 6:54.
6. *Statues at Large* (Cambridge: Bentham, 1765), 16:102–3.
7. Albion, *Forests and Sea Power*, 261.
8. Vietze, *White Pine*, 75.
9. Joseph Thing, deposition, Exeter, NH, May 13, 1734.
10. John Lusken, deposition, Exeter, NH, May 13, 1734.
11. David Dunbar, letter to Willian Belcher, in *Calendar of State Papers*, colonial ser. (London: Her Majesty's Stationery Office, 1953), 41:144.
12. Town of Exeter, NH, petition to William Belcher, May 6, 1734, 2.
13. *Calendar of State Papers*, 41:92.
14. Henry Dawson, *The Historical Magazine and Notes and Queries Concerning the Antiquities, History, and Biography of America*, ser. 2 (Morrisania, NY, 1870), 8:14.
15. Dawson, *The Historical Magazine*, 8:14
16. Dawson, *The Historical Magazine*, 8:15.
17. Dawson, *The Historical Magazine*, 8:14.
18. Dawson, *The Historical Magazine*, 8:15.
19. Dawson, *The Historical Magazine*, 8:15.
20. Dawson, *The Historical Magazine*, 8:14.
21. *Calendar of State Papers*, 41:92.
22. *Calendar of State Papers*, 41: 93.
23. Strother Roberts, "Pines, Profits, and Politics: Responses to the White Pine Acts in the Colonial Connecticut River Valley," *New England Quarterly* 83 (2010): 73–101.

24. Benning Wentworth, letter to Roger Wolcott, in *Collections of the Connecticut Historical Society* (Hartford: Connecticut Historical Society, 1918), 17:1.

25. Response to Blake petition, *Journal of the Commissioners for Trade and Plantation from January 1759 to December 1763* (London: His Majesty's Stationery Office, 1935), 28.

26. *New Hampshire Gazette*, February 21, 1772.

27. William Little, *The History of Weare, New Hampshire, 1735–1888* (Lowell, MA: Huse, 1888), 187.

28. *New Hampshire Gazette*, April 24, 1772.

29. Little, *History of Weare*, 190.

30. Little, *History of Weare*, 191.

Chapter 3: Boundary Disputes and the Aroostook War

1. John Fairfield, letter to John Harvey, February 19, 1839, in *Correspondence Relating to the Boundary between British Possessions in North America and the United States under the Treaty of 1783* (London: Harrison, 1840), 39–40.

2. Fairfield, letter to Harvey, February 19, 1839, 67.

3. Hunter Miller, ed., *Treaties and Other International Acts of the United States of America* (Washington, DC: U.S. Government Printing Office, 1931), 2:430.

4. Miller, *Treaties and Other International Acts*, 2:574.

5. Francis Carroll, *A Good and Wise Measure* (Toronto: University of Toronto Press, 2001).

6. Carroll, *A Good and Wise Measure*, 2:63.

7. Pierre Du Perreé, letter to Thomas Wetmore, February 20, 1819, in *American State Papers: Documents, Legislative and Executive, of the Congress of the United States* (Washington, DC: Gales and Seaton, 1859), 6:849.

8. Charles Bagot, letter to Thomas Barclay, December 8, 1818, *American State Papers*, 6:849.

9. For a thorough account of the Aroostook War, see W. E. Campbell, *The Aroostook War of 1839* (Fredericton, NB: Goose Lane, 2013).

10. Campbell, *The Aroostook War*, 31.

11. John Fairfield, message to the Maine Senate and House of Representatives, January 23, 1839, *Resolves of the Nineteenth Legislature of the State of Maine* (Augusta, ME: Smith and Robinson, 1839), 149.

12. Campbell, *The Aroostook War*, 59.

13. *Portland Advertiser*, February 26, 1839.

14. John Harvey, proclamation, February 13, 1839, in *Correspondence Relating to the Boundary*, 19.

15. John Harvey, letter to John Fairfield, February 13, 1839, in *Correspondence Relating to the Boundary*, 38.

16. John Fairfield, letter to John Harvey, February 19, 1839, in *Correspondence Relating to the Boundary*, 39–40.

17. Fairfield, letter to Harvey, February 19, 1839.

18. Frederick Street, letter to Charles Jarvis, February 17, 1839, in *Correspondence Relating to the Boundary*, 45.

19. Charles Jarvis, letter to Frederick Street, February 17, 1839, in *Correspondence Relating to the Boundary*, 45–46.

20. Henry Fox, letter to John Forsyth, February 23, 1839, in *Correspondence Relating to the Boundary*, 20.

21. John Forsythe, letter to Henry Fox, February 25, 1839, in *Correspondence Relating to the Boundary*, 24.

22. Fox-Forsyth memorandum, February 27, 1839, in *Correspondence Relating to the Boundary*, 26–27.

23. John Harvey, letter to Henry Fox, March 6, 1839, in *Correspondence Relating to the Boundary*, 78.

24. Winfield Scott, *Memoirs of Lieutenant General Scott, LL.D* (New York: Sheldon and Company, 1864), 2:336–38, 344–46.

25. Winfield Scott, John Harvey, and John Forsyth, declaration, March 21, 1839, in *Correspondence Relating to the Boundary*, 81–82.

26. Scott et al., declaration.

27. Alec McEwen, *In Search of the Highlands: Mapping the Canada-Maine Boundary. The Journals of Featherstonhaugh and Mudge, August to November 1839* (Fredericton, NB: Acadiensis, 1988), 33, 35.

28. Edward Kent, address to the Maine Senate and House of Representatives, January 15, 1841, in *Resolves of the Twenty-First Legislature of the State of Maine* (Augusta, ME: Severance and Dorr, 1841), 666.

29. Carroll, *A Good and Wise Measure*, 266–76.

30. George Ticknor Curtis, *Life of Danial Webster* (New York: Appleton and Company, 1870), 2:113.

Chapter 4: White Pines and Red Spruce in the Nineteenth-Century Maine Woods

1. John Springer, *Forest Life and Forest Trees* (New York: Harper and Brothers, 1851), 43.

2. Springer, *Forest Life and Forest Trees*, 52.

3. Springer, *Forest Life and Forest Trees*, 53.

4. Springer, *Forest Life and Forest Trees*, 71.

5. Springer, *Forest Life and Forest Trees*, 71.

6. Springer, *Forest Life and Forest Trees*, 71.

7. Springer, *Forest Life and Forest Trees*, 83.

8. Springer, *Forest Life and Forest Trees*, 165–67.

9. Springer, *Forest Life and Forest Trees*, 162–63.

10. William Williamson, *History of the State of Maine*, vol. 2 (Hallowell, ME: Glazier, Masters, and Smith, 1839).

11. John Godfrey, "The Ancient Penobscot," *Collections of the Maine Historical Society* 7 (1876): 4.

12. Peter Force, ed., *American Archives*, ser. 4 (Washington, DC: Clarke and Force), 2:1432.

13. Force, *American Archives*, 2:1433.

14. Frederick Allis, *William Bingham's Maine Lands, 1790–1820* (Boston: Colonial Society of Massachusetts, 1954).

15. Allis, *William Bingham's Maine Lands*, 653.

16. *Acts and Laws of the Commonwealth of Massachusetts*, 1796–97 (Boston: Adams and Larkin, 1896), 569–70.

17. Vine Deloria Jr. and Raymond DeMallie, eds., *Documents of American Indian Diplomacy* (Norman: University of Oklahoma Press, 1999), 2:1094.

18. Deloria and DeMallie, *Documents of American Indian Diplomacy*, 2:1094.

19. *Resolves of the General Court of the Commonwealth of Massachusetts* (Boston: Young and Minns, 1806), chap. 27 [June 1803].

20. Micha Pawing, ed., *Wabanaki Homeland and the New State of Maine: The 1820 Journal and Plans of Survey of Joseph Treat* (Amherst: University of Massachusetts Press, 2017).

21. *Acts and Resolves of the Twenty-Second Legislature of the State of Maine* (Augusta, ME: Smith, 1842), 253–55.

22. *Acts and Resolves of the Twenty-Second Legislature*, 258–60.

23. *Acts and Resolves of the Twenty-Second Legislature*, 262–63.

24. Donna Loring, Eric Mehnert, and Joseph Gousse, *One Nation, Under Fraud: A Remonstrance* (Augusta, ME: Permanent Commission on the Status of Racial, Indigenous, and Tribal Populations, 2022), 13.

25. Mark Trafton, "Remonstrance of the Penobscot Tribe (1833)," Digital Maine Repository, https://digitalmaine.com.

26. *Eastern Argus* (Portland, Maine), February 10, 1834.

27. Williamson, *History*, 670.

28. Hugh McCulloch, *Men and Measures of Half a Century* (New York: Scribner, 1889), 215.

29. Stewart Holbrook, *Holy Old Mackinaw* (New York: Macmillan, 1938), 14.

30. Henry David Thoreau, *The Maine Woods* (Boston: Ticknor and Fields, 1864), 19, 83.

31. For a thorough account of Chamberlain Lake history, see Lew Dietz, *The Allagash* (New York: Holt, Rinehart, and Winston, 1968); and Dean Bennett, *The Wilderness from Chamberlain Farm* (Washington, DC: Island Press, 2001).

32. *Acts and Resolves Passed by the Twenty-Sixth Legislature of the State of Maine* (Augusta, ME: Johnson, 1846), chap. 386, 488–90.

33. *Acts and Resolves Passed by the Twenty-Sixth Legislature*, chap. 387, 490–93.

34. William Parrott, "Report of William P. Parrott," in *Documents Printed by the Order of the Legislature of the State of Maine* (Augusta, ME: Smith, 1844), 23–29.

35. Parrott, "Report," 23–29; Levi Bradley, "Report of the Land Agent of the State of Maine," in *Documents Printed by the Order of the Legislature of the State of Maine* (Augusta, ME: Smith, 1844), 10–11.

36. *Acts and Resolves Passed by the Twenty-Sixth Legislature*, chap. 389, 494–95.

37. Richard Judd, *Aroostook: A Century of Logging in Northern Maine* (Orono: University of Maine Press, 1989).

38. *Acts and Resolves Passed by the Thirty-First Legislature of the State of Maine* (Augusta, ME: Johnson, 1852), chap. 525, 519–20.

39. Thoreau, *The Maine Woods*, 235.

40. Thoreau, *The Maine Woods*, 244.

41. *Acts and Resolves Passed by the Thirty-Ninth Legislature of the State of Maine* (Augusta, ME: Stevens and Sayward, 1860), chap. 464, 425.

42. Ezekiel Holmes, "Preliminary Report upon the Natural History and Geology of the State of Maine," in *Sixth Annual Report of the Secretary of the Maine Board of Agricultural* (Augusta, ME: Stevens and Sayward, 1861), 348.

43. Walter Wells, *The Water-Power of Maine* (Augusta, ME: Sprague, Owen, and Nash, 1869), 112, 459.

44. Lucius Hubbard, *Woods and Lakes of Maine: A Trip from Moosehead Lake to New Brunswick* (Boston: Ticknor, 1883), 78.

45. L. P. Wyman, *The Golden Boys Save the Chamberlain Dam* (New York: Burt, 1927).

46. Andrew M. Barton, Alan S. White, and Charles V. Cogbill, *The Changing Nature of the Maine Woods* (Durham: University of New Hampshire Press, 2012), 115.

47. Robert Pike, *Tall Trees, Tough Men* (New York: Norton, 1999).

48. Barton et al., *Changing Nature*, 121.

49. Barton et al., *Changing Nature*, 121.

50. Alvin Lombard, U.S. Patent 674,737, issued May 21, 1901.

51. Pike, *Tall Trees*; Prescott Howard, "The Era of the Lombard Log Hauler," *Forest History Newsletter* 6 (1962): 2–8.

52. O. A. Harkness, "The Eagle Lake Tramway," *Northern Logger* 14 (1966): 45–46

53. Pike, *Tall Trees*; Bennett, *The Wilderness*.

Chapter 5: Terpenes and Their Part in Christmas, Spruce Beer, and Tree Defenses

1. Dallas Lore Sharp, "The Woods of Maine," *Harper's*, June 1, 1922, 185.

2. Alan Crozier, Michael N. Clifford, and Hiroshi Ashihara, *Plant*

Secondary Metabolites: Occurrence, Structure and Role in Human Diet (Ames, IA: Blackwell, 2006), 55–71.

3. Ernst Von Rudloff, "Volatile Leaf Oil Analysis in Chemosystematic Studies of North American Conifers," *Biochemical Systematics and Ecology* 2 (1975): 131–67.

4. Clarence Day, *Farming in Maine: 1860–1940* (Orono: University of Maine Press, 1963).

5. William P. Atherton, "Notes by the Way," *Maine Farmer*, August 21, 1890.

6. Josiah McIntyre, "Typical Abandoned Farm," *Maine Farmer*, October 29, 1891.

7. Albert Pease, "Abandoned Farms," *Maine Farmer*, July 19, 1894.

8. "Abandoned Farms Again," *Maine Farmer*, October 1, 1891.

9. "Abandoned Farms Again."

10. S. F. Emerson, "Abandoned Farms," *Maine Farmer*, February 12, 1891.

11. Emerson, "Abandoned Farms."

12. Aunt Patience, "Keeping the Boys and Girls on the Farm," *Maine Farmer*, September 10, 1891.

13. "Deserted Farms," *Maine Farmer*, January 28, 1897.

14. Atherton, *"Noes by the Way."*

15. Robert Frost, "Christmas Trees," *Mountain Interval* (New York: Holt, 1916), 11–14.

16. Thomas McAdam, "Christmas Greens and Flowers," *Country Life in America* 5 (December 1903): 136–37.

17. Austin Wilkins, *The Forests of Maine: Their Extent, Character, Ownership, and Products* (Augusta: Maine Forest Service, 1932), 80.

18. Federal Reserve Bank of Boston, "New England's Christmas Tree Industry: A Declining Past—A Hopeful Future," *New England Business Review* 38 (December 1956): 2–3.

19. Henry David Thoreau, *The Maine Woods* (Boston: Ticknor and Fields, 1864), 26.

20. Pehr Kalm, "Pehr Kalm's Description of Spruce Beer," trans. Ester Louise Larsen, *Agricultural History* 22 (1948): 142–43.

21. Kalm, "Pehr Kalm's Description of Spruce Beer," 142–43.

22. Jonathan Pereira, *The Elements of Materia Medica and Therapeutics* (Philadelphia: Lea and Blanchard, 1843), 2:184.

23. Crozier et al., *Plant Secondary Metabolites.*

24. Theodore Gerrish, *Army Life; A Private's Reminiscences of the Civil War* (Portland, ME: Hoyt, Fogg, and Donham, 1882), 13.

25. Wilkins, *The Forests of Maine*, 94; John Gordon Dorrance, *The Story of the Forest* (New York: American Book Company, 1916), 203–5.

26. Charles A. J. Farrar, *Through the Wilds* (Boston: Estes and Lauriat, 1892), 374.

27. "Overseer's Report," *Harvard Lampoon*, May 23, 1888, 102–3.

28. Mary Rogers Miller, *Outdoor Work* (New York: Doubleday, Page, 1916), 82.

29. Wilkins, *The Forests of Maine*, 95.

30. "Deserted Farms."

31. J. W. Hudgins, Erik Christiansen, and Vincent Franceschi, "Induction of Anatomically Based Defense Responses in Stems of Diverse Conifers by Methyl Jasmonate: A Phylogenetic Perspective," *Tree Physiology* 24 (2004): 251–64; Paul Krokene, Nina Elisabeth Nagy, and Trygve Krekling, "Traumatic Resin Ducts and Polyhphenolic Parenchyma Cells in Conifers," in *Induced Plant Resistance to Herbivory*, ed. Andreas Schaller (New York: Springer Science, 2008), 147–69.

32. Michael Philips and Rodney Croteau, "Resin-Based Defenses in Conifers," *Trends in Plant Science* 4 (1999): 184–90.

33. Jonathan Gershenzon, "Metabolic Costs of Terpenoid Accumulation in Higher Plants," *Journal of Chemical Ecology* 20 (1994): 1281–1328

34. Elizabeth Tomlin, Eva Antonejevic, Rene Alfaro, and John Borden, "Changes in Volatile Terpene and Diterpene Resin Acid Composition of Resistant and Susceptible White Spruce Leaders Exposed to Simulated White Pine Weevil Damage," *Tree Physiology* 20 (2000): 1087–95; Kimberly Wallin and Kenneth Raffa, "Altered Constitutive and Inducible Phloem Monoterpenes Following Natural Defoliation of Jack Pine: Implications to Host Mediated Interguild Interactions and Plant Defense Theory," *Journal of Chemical Ecology* 25 (1999): 861–80; Diane Martin, Dorothea Tholl, Jonathan Gershenzon, and Jorg Bohlmann, "Methyl Jasmonate Incudes Traumatic Resin Ducts, Terpenoid Biosynthesis, and Terpenoid Accumulation in Developing Xylem of Norway Spruce Stems," *Plant Physiology* 129 (2002): 1003–18.

35. Lulu Dai, Haiming Gao, and Hui Chen, "Expression Levels of Detoxification Enzyme Genes from *Dendroctonus armandi* (Coleoptera: Curculionidaea) Fed on a Solid Diet Containing Pine Phloem and Terpenoids," *Insects* 12 (2021): 926–41.

36. Celia Boone, Ken Keefover-Ring, Abigail Mapes, Aaron Adams, Jorg Bohlmann, and Kenneth Raffa, "Bacteria Associated with a Tree-Killing Insect Reduce Concentrations of Plant Defense Compounds," *Journal of Chemical Ecology* 39 (2013): 1003–6.

37. Philips and Croteau, "Resin-based Defenses," 185.

38. Jean-Claude Gregoire, Daniel Coullien, Ralph Krebber, Winfried Konig, Holger Meyer, and Wittko Francke. "Orientation of *Rhizophagus grandis* (Coleoptera: Rhizophagidae) to Oxygenated Monoterpenes in a Species-Specific Predator-Prey Relationship," *Chemoecology* 3 (1992): 14–18.

39. Stephen Sondheim and Hugh Wheeler, *Sweeney Todd: The Demon Barber of Fleet Street* (New York: Dodd, Mead, 1979), 105.

40. Miller, *Outdoor Work*, 80–81.

Chapter 6: Balsam Fir Resins in Medicine, Microscopes, and Germ Theory

1. Alexander Mackenzie, *Voyages from Montreal, on the River St. Laurence, through the Continent of North-America, to the Frozen and Pacific Oceans, in the Years 1789 and 1793* (New York: Evert Duyckinck, 1803), 215.
2. Mackenzie, *Voyages from Montreal*, 217–18.
3. Héloïse Coté, Marie-Anne Boucher, André Pichette, Benoit Roger, and Jean Legault, "New Antibacterial Hydrophobic Assay Reveals *Abies balsamea* Oleoresin Activity against *Staphylococcus aureus* and MRSA," *Journal of Ethnopharmacology* 194 (2016): 684–89.
4. Huron Smith, "Ethnobotany of the Forest Potawatomi Indians," *Bulletin of the Public Museum of the City of Milwaukee* 7 (1933): 69.
5. Coté et al., "New Antibacterial Hydrophobic Assay."
6. John Gunn, *Gunn's New Domestic Physician* (Cincinnati, OH: Moore, Wilstach, and Keys, 1861), 5.
7. John Kost, *Elements of Materia Medica and Therapeutics* (Cincinnati, OH: Moore, Wilstach, and Keys, 1858), 437.
8. Gunn, *Gunn's New Domestic Physician*, 758.
9. William Lewis, *The New Dispensatory* (London: Wingrave, 1799), 252.
10. Lewis, *The New Dispensatory*, 252.
11. Morris Mattson, *The American Vegetable Practice* (Boston: Hale, 1841), 269.
12. Kost, *Elements of Materia Medica*, 438.
13. Kost, *Elements of Materia Medica*, 438.
14. Gunn, *Gunn's New Domestic Physician*, 759; Edward Goodman Clarke, *A Conspectus of the London, Edinburgh, and Dublin Pharmacopoeias* (London: Cox and Son, 1814), 26.
15. Mattson, *The American Vegetable Practice*, 268.
16. Savile Bradbury, *The Evolution of the Microscope* (London: Pergamon, 1967).
17. Joseph Jackson Lister, "On Some Properties in Achromatic Object-Glass Applicable to the Improvement of the Microscope," *Philosophical Transactions of the Royal Society of London* 120 (1830): 194.
18. Lister, "On Some Properties," 193.
19. Brian Bracegirdle, "J. J. Lister and the Establishment of Histology," *Medical History* 21 (1977): 187.
20. Joseph Lister, "Obituary Notice of the Late Joseph Jackson Lister," *Monthly Microscopical Journal* 3 (1870): 134.
21. Henry Smith Williams, "The Century's Progress in Anatomy and Physiology," *Harper's* 96 (March 1898): 624.
22. Andrew Pritchard, *A List of Two Thousand Microscopic Objects* (London: Whittaker, 1835), 6.

23. Andrew Pritchard and Charles R. Goring, *Microscopic Illustrations of Living Objects* (London: Whittaker, 1838), 77, 236.
24. John W. Griffith, "On the Different Modes of Preserving Microscopic Objects," *Annals and Magazine of Natural History* 12 (1843): 114–15.
25. Griffith, "On the Different Modes," 113.
26. William B. Carpenter, *The Microscope and Its Revelations* (London: Churchill, 1856), 224.
27. William B. Carpenter, *The Microscope and Its Revelations*, rev. ed. (London: Churchill, 1868), 214–16.
28. Carpenter, *The Microscope and Its Revelations*, rev. ed., 214.
29. "Catch-Pennies," *Household Words*, August 19, 1854, 5–6.
30. "Catch-Pennies," 5–6.
31. "Catch-Pennies," 5–6.
32. Andrew Pritchard, *A History of Infusorial Animalcules* (London: Whittaker, 1852), 2.
33. Patrice Debré, *Louis Pasteur*, trans. Elborg Forster (Baltimore: Johns Hopkins University Press, 1994).
34. Lindsey Fitzharris, *The Butchering Art* (New York: Scientific American and Farrar, Strauss, and Giroux, 2017).
35. Joseph Lister, "On a New Method of Treating Compound Fracture, Abscess, Etc. with Observations on the Condition of Suppuration," *Lancet*, March 16, 1867, 327.
36. Lister, "On a New Method," 327.
37. "Joseph Jackson Lister and the Achromatic Microscope," *British Medical Journal*, May 28, 1898, 1407.
38. Lister, "On a New Method," 327.

Chapter 7: Pitch Pine Resins in Medicine, Naval Stores, and Colonial Conflicts

1. *New Hampshire Gazette*, February 18, 1774.
2. Silas Little, "Prescribed Burning as a Tool of Forest Management in the Northeastern States," *Journal of Forestry* 51 (1953): 496.
3. John Winthrop Jr., "Of the Manner of Making Tar and Pitch in New England," in *History of the Royal Society of London for Improving of Natural Knowledge from Its First Rise*, vol. 1, ed. Thomas Birch (London: Millar, 1756), 101.
4. Jeffery P. Brain, *Fort St. George: Archeological Investigation of the 1607–1608 Popham Colony* (Augusta: Maine Archeological Society, 2007).
5. Lawrence Earley, *Looking for Longleaf* (Chapel Hill: University of North Carolina Press, 2004), 99–101; Strother Roberts, "The Fight for a New England Turpentine Trade: Empires, Markets, and the Colonial Landscape at the Turn of the Eighteenth Century," *New England Quarterly* 92 (2019): 395–96.
6. John C. Gunn, *Gunn's New Domestic Physician* (Cincinnati, OH: Moore, Wilstach, and Keys, 1861), 406, 547, 645, 647, 665.

7. Gunn, *Gunn's New Domestic Physician*, 702.
8. Charles Fuller, *Personal Recollections of the War of 1861* (Sherburne, NY: News Job Printing House, 1906), 103.
9. Robert Christison, *A Dispensatory, or Commentary on the Pharmaco-poeias of Great Britain* (Edinburgh: Black, 1842), 922; Samuel O. L. Potter, *Handbook of Materia Medica, Pharmacy, and Therapeutics* (Philadelphia: Blakiston, 1887), 376.
10. Gunn, *Gunn's New Domestic Physician*, 325, 636, 727; Christison, *A Dispensatory*, 922.
11. Christison, *A Dispensatory*, 922.
12. Gunn, *Gunn's New Domestic Physician*, 335.
13. Jonathan Pereira, *The Elements of Materia Medica and Therapeutics* (Philadelphia: Lea and Blanchard, 1843), 2:314.
14. Roberts, "The Fight for a New England Turpentine Trade,"401–2; Mikko Airaksinen, "Tar Production in Colonial America," *Environment and History* 2 (1996): 120
15. Airaksinen, "Tar Production in Colonial America," 120.
16. Airaksinen, "Tar Production in Colonial America," 115.
17. Airaksinen, "Tar Production in Colonial America," 116; Thomas Gamble, ed., *Naval Stores: History, Production, Distribution, and Consumption* (Savannah, GA: Review Publishing and Printing, 1921), 43.
18. Gamble, *Naval Stores*, 16; Earley, *Looking for Longleaf*, 93–94.
19. Earley, *Looking for Longleaf*, 87, 89.
20. Robert Hay and Vincent McInerney, eds., *Landsman Hay* (Barnsley, UK: Seaforth, 2010), 153–54.
21. Earley, *Looking for Longleaf*, 87, 89–90.
22. Thomas Morton, *New English Canaan* (Amsterdam: Stam, 1637), 63.
23. Winthrop, "Of the Manner of Making Tar," 99.
24. Winthrop, "Of the Manner of Making Tar," 99–101.
25. Roberts, "The Fight for a New England Turpentine Trade," 403–4.
26. Roberts, "The Fight for a New England Turpentine Trade," 423–24.
27. Clarence W. Bowen, *The Boundary Disputes of Connecticut* (Boston: Osgood, 1882), 19.
28. Roberts, "The Fight for a New England Turpentine Trade," 417.
29. Roberts, "The Fight for a New England Turpentine Trade," 418.
30. Roberts, "The Fight for a New England Turpentine Trade," 408.
31. Charles J. Hoadly, *Public Records of the Colony of Connecticut from August, 1689 to May, 1706* (Hartford, CT: Case, Lockwood, and Brainard, 1868), 4:301.
32. Council of the Massachusetts Bay, minutes, April 1, 1703, in *Calendar of State Papers, Colonial Series, America and West Indies, 1702–1703* (London: Her Majesty's Stationery Office, 1913), 309.
33. Hoadly, *Public Records*, 443–44.
34. Joseph Sheldon and Jonathan Remmington [*sic*], petition, May

31, 1704, in *Acts and Resolves, Public and Private, of the Province of Massachusetts* (Boston: Wright and Potter, 1895), 8:540.

35. "Chapter 82: Resolve for Allowing and Paying Nineteen Pounds out the Province Treasury to Joseph Sheldin and Jonathan Remington [*sic*]," *Acts and Resolves*, 8:148.

36. Robert Outland, *Tapping the Pines: The Naval Stores Industry in the American South* (Baton Rouge: Louisiana State University Press, 2004).

37. Airaksinen, "Tar Production in Colonial America," 118.

38. Eleanor L. Lord, *Industrial Experiments in the British Colonies of North America* (Baltimore: Johns Hopkins University Press, 1898), 142.

39. *Statues at Large* (Cambridge: Bentham, 1764), 11:109

40. Lord, *Industrial Experiments*, 142.

41. Roger Wolcott, letter to Josiah Willard, December 17, 1750, in *Collections of the Connecticut Historical Society* (Hartford: Connecticut Historical Society, 1916), 16:19.

42. Roberts, "The Fight for a New England Turpentine Trade," 425.

43. Bowen, *The Boundary Disputes*, 58–59.

44. Airaksinen, "Tar Production in Colonial America," 119.

45. Benjamin J. Irvin, "Tar, Feathers, and the Enemies of American Liberties, 1768–1776," *New England Quarterly* 76 (2003): 204–6.

46. Irvin, "Tar, Feathers, and the Enemies of American Liberties," 201–2.

47. Irvin, "Tar, Feathers, and the Enemies of American Liberties," 205.

48. Irvin, "Tar, Feathers, and the Enemies of American Liberties," 208–9.

49. Irvin, "Tar, Feathers, and the Enemies of American Liberties," 214–17.

50. Emanuel Hertz, *Lincoln Talks: A Biography in Anecdote* (New York: Halcyon, 1941), 259.

Chapter 8: Hemlock Tannins and Making Leather

1. Ulysses S. Grant, *Personal Memoirs of U. S. Grant* (New York: Webster, 1885), 26.

2. Alan Crozier, Michael N. Clifford, and Hiroshi Ashihara, *Plant Secondary Metabolites: Occurrence, Structure, and Role in Human Diet* (Ames, IA: Blackwell, 2006), 1–24.

3. Raymond V. Barbehenn and C. Peter Constabel, "Tannins in Plant-Herbivore Interactions," *Phytochemistry* [Oxford University] 72 (2011): 1552.

4. Barbehenn and Constabel, "Tannins in Plant-Herbivore Interactions," 1555.

5. Barbehenn and Constabel, "Tannins in Plant-Herbivore Interactions," 1558; Martha R. Bajec and Gary J. Pickering, "Astringency: Mechanisms and Perception," *Critical Reviews in Food Science and Nutrition* 48 (2008): 858–75.

6. H. L. Gibbins and G. H. Carpenter, "Alternative Mechanisms of Astringency—What Is the Role of Saliva?," *Journal of Texture Studies* 44 (2013): 364–75.

7. Sarah Ployon, Martine Morzel, Christine Belloir, Aline Bonnette, Eric Bourillot, Loic Briand, et al., "Mechanisms of Astringency: Structural Alteration of the Oral Mucosal Pellicle by Dietary Tannins and Protective Effect of bRPRs," *Food Chemistry* 253 (2018): 79–87.

8. Anthony D. Covington, "Modern Tannin Chemistry," *Chemical Society Reviews* 26 (1997): 111–26.

9. William Shakespeare, *The Tragedy of Hamlet, Prince of Denmark* (c. 1600; New York: French and Son, 1866), 30

10. Steps in traditional leather making are described in Peter C. Welsh, "A Craft That Resisted Change: American Tanning Practices to 1850," *Technology and Culture* 4 (1963): 299–317; and in Andrew Ure, *A Dictionary of Arts, Manufactures, and Mines* (New York: Appleton, 1858), 2:54–58.

11. Samuel Matthews, *Tenth Annual Report of the Bureau of Industrial and Labor Statistics for the State of Maine, 1896* (Augusta, ME: Burleigh and Flynt, 1897), 56.

12. Jacques Ferland, "The Command of Money in Shaws' Borderlands, 1859–1887," in *New England and Maritime Provinces: Connections and Comparisons*, ed. Stephen J. Hornsby and John G. Reid (Montreal: McGill-Queen's University Press, 2005), 164–65.

13. Matthews, *Tenth Annual Report*, 64–65; Minnie Atkinson, *Hinckley Township or Grand Stream Lake Plantation, A Sketch* (Newburyport, MA: Newburyport Herald Press, 1920), 58.

14. Atkinson, *Hinckley Township*, 33; Amos Wilder, "Land Locked Salmon," *Kennebec Journal*, June 2, 1883.

15. Matthews, *Tenth Annual Report*, 67.

16. Wilder, "Land Locked Salmon."

17. Ferland, "The Command of Money," 165.

18. Martin Butler, *Maple Leaves and Hemlock Branches* (Fredericton, NB: Gleaner Job Office, 1889), 42.

19. Butler, *Maple Leaves and Hemlock Branches*, 42; Matthews, *Tenth Annual Report*, 72–76.

20. Matthews, *Tenth Annual Report*, 72–76.

21. Matthews, *Tenth Annual Report*, 57, 72.

22. Ferland, "The Command of Money," 167–69.

23. "Suspensions in the Boot and Shoe and Leather Trades," *Boston Daily Advertiser*, July 31, 1883.

24. "Large Failure," *Maine Farmer*, August 2, 1883, 2.

25. Ferland, "The Command of Money," 170, 172.

26. Atkinson, *Hinckley Township*, 56, 58

27. Atkinson, *Hinckley Township*, 60–61.

Appendix: Finding and Identifying Conifers at Acadia National Park

1. All of the species illustrations are modified from Charles S. Sargent, *Silva of North America*, vols. 10–12 (Boston: Houghton Mifflin, 1896–98).
2. Barrington Moore and Norman Taylor, "Vegetation of Mount Desert Island, Maine, and Its Environment," *Brooklyn Botanic Garden Memoirs* 3 (1927): 1–151.
3. Duncan Johnson and Alexander Skutch. "Littoral Vegetation on a Headland of Mt. Desert Island Maine. III. Adlittoral or Non-submersible Region," *Ecology* 9 (1928): 429–48.
4. George Dorr, *The Acadian Forest* (Bar Harbor, ME: Wild Gardens of Acadia, 1922), 6.
5. Thomas Ledig, Peter Smouse, and John Hom, "Postglacial Migration and Adaptation for Dispersal in Pitch Pine (Pinacaea)," *American Journal of Botany* 102 (2015): 2074–91.
6. Michael Greenwood, William Livingston, Michael Day, Shawn Kenaley, Alan White, and John Brissette, "Contrasting Modes of Survival by Jack and Pitch Pine at a Common Range Limit," *Canadian Journal of Forest Research* 32 (2002): 1662–74.
7. Henry David Thoreau, *Walden* (Boston: Osgood, 1878), 41.
8. Nathan Havill, Ligia Viera, and Scott Salom, *Biology and Control of the Hemlock Wooly Adelgid* (FHTET-2014-05) (Washington, DC: U.S. Department of Agriculture, Forest Health Technology Enterprise Team, 2016).

Index

Page numbers followed by an "f" indicate figures. Those followed by a "t" indicate tables.